Name ______________________ Class ______________ Date __

Skills Worksheet

Directed Reading

AF769478

Section: Neurons and Nerve Impulses

In the space provided, write the letter of the description that best matches the term or phrase.

______ **1.** neurons	**a.** receives information from other neurons
______ **2.** dendrite	**b.** a long extension of the cytoplasm that conducts nerve impulses
______ **3.** axon	**c.** a bundle of axons
______ **4.** nerve	**d.** specialized cells that transmit information throughout the body
______ **5.** myelin sheath	**e.** layer of insulation on the axons of some neurons

Read each question, and write your answer in the space provided.

6. What causes the membrane potential of a neuron?

____

____

____

____

____

7. Why is the resting potential of a typical neuron negative?

____

____

____

____

____

____

____

____

8. What is an action potential?

____

____

____

____

Study the following steps in the conduction of a nerve impulse. Determine the order in which the steps take place. Write the number of each step in the space provided.

______ **9.** The resting potential is fully restored.

______ **10.** Voltage-gated potassium channels open, causing potassium ions to flow out of the axon.

______ **11.** The neuron is at the resting potential of about −70 mV.

______ **12.** A stimulus causes voltage-gated sodium channels to open, and sodium ions rapidly flow into the axon. This causes the membrane potential to increase to about +40 mV.

Study the following steps in the synaptic transmission of a nerve impulse from a presynaptic neuron to a postsynaptic cell. Determine the order in which the steps take place. Write the number of each step in the space provided.

______ **13.** Vesicles in the presynaptic neuron fuse with the cell membrane.

______ **14.** Neurotransmitter molecules are released into the synaptic cleft.

______ **15.** A nerve impulse arrives at an axon terminal of the presynaptic neuron.

______ **16.** Neurotransmitters bind to receptor proteins on the postsynaptic cell.

Name ______________________ Class ______________ Date ____________

Skills Worksheet

Directed Reading

Section: Structures of the Nervous System

In the space provided, explain how the terms in each pair differ in meaning.

1. central nervous system, peripheral nervous system

____

____

____

____

____

2. sensory neurons, motor neurons

____

____

____

____

____

In the space provided, write the letter of the description that best matches the term or phrase.

______ **3.** brain	**a.** helps regulate breathing and heart rate
______ **4.** cerebrum	**b.** a sudden, involuntary contraction of muscles in response to a stimulus
______ **5.** cerebellum	**c.** the site of learning, memory, perception, and intellectual function
______ **6.** reflex	**d.** relays sensory information to the cerebral cortex
______ **7.** thalamus	**e.** the dense cable of nervous tissue that runs through the vertebral column
______ **8.** hypothalamus	**f.** regulates balance, posture, and movement
______ **9.** spinal cord	**g.** the body's main processing center

Complete each statement by writing the correct term or phrase in the space provided.

10. Peripheral motor neurons that stimulate skeletal muscles are part of the

________________________ nervous system.

11. The ________________________ division of the autonomic nervous system keeps your body functioning even when you are asleep.

12. The ________________________ division of the autonomic nervous system dominates in times of physical or emotional stress.

Name ______________________ Class ______________ Date ____________

Skills Worksheet

Directed Reading

Section: Sensory Systems

Read each question, and write your answer in the space provided.

1. What is a sensory receptor?

____

____

____

2. What types of sensory receptors are located in the skin?

____

____

____

3. Why is pain an important sensation?

____

____

4. Which type of sensory receptor responds to the distortion or bending of tissue?

____

____

5. What are the four lobes of the cerebral cortex?

____

____

____

Complete each statement by writing the correct term or phrase in the space provided.

6. The ______________________ is a lining of photoreceptors and neurons on the back inner surface of the eye.

7. The photoreceptors called ______________________ respond best to dim light, and those called ______________________ enable color vision.

8. The ______________ ______________ is made up of the axons of the neurons in the retina.

Read each question, and write your answer in the space provided.

9. How does the cochlea of the inner ear help you hear?

____

____

____

10. How do the semicircular canals help you keep your balance?

____

____

____

Complete each statement by writing the correct term or phrase in the space provided.

11. Taste cells detect sugars, acids, alkaloids, and ______________ .

12. Chemoreceptors that detect odors are called ______________ receptors.

Name ______________________ Class ______________ Date ____________

Skills Worksheet

Directed Reading

Section: Drugs and the Nervous System

Read each question, and write your answer in the space provided.

1. What is a psychoactive drug?

____

____

____

2. Which type of psychoactive drug puts the user at risk of drowsiness or depression?

____

____

____

3. What are the psychoactive effects of inhalants?

____

____

____

4. Give two examples of psychoactive drugs that contain THC, and describe their psychoactive effects.

____

____

____

5. What risks are associated with the use of narcotics?

____

____

____

In the space provided, write the letter of the description that best matches the term or phrase.

_____ **6.** addiction

_____ **7.** tolerance

_____ **8.** withdrawal

_____ **9.** stimulant

a. a drug that increases the activity of the central nervous system

b. increasing amounts of a drug are needed to achieve the desired sensation

c. a physiological response that alters normal functioning of neurons and synapses

d. symptoms caused by removing an addictive drug from the body

Read each question, and write your answer in the space provided.

10. How does cocaine affect dopamine-producing neurons?

____

____

____

____

11. Why is cocaine addictive?

____

____

____

____

12. What is a depressant?

____

____

13. What are some of the harmful effects of alcoholism?

____

____

____

Complete each statement by writing the correct term or phrase in the space provided.

14. The highly addictive stimulant found in the leaves of the tobacco plant is called ______________________ .

15. In the brain, nicotine mimics the action of the neurotransmitter ______________________ .

16. Inhaled smoke contains hundreds of toxic chemicals called ______________________ .

17. After reaching the spinal cord, a pain signal is suppressed by neurotransmitters called ______________________ .

18. Codeine, morphine, and heroin are all derived from a substance called ______________________ .

Name ______________________ Class ______________ Date ____________

Skills Worksheet

Active Reading

Section: Neurons and Nerve Impulses

Read the passage below. Then answer the questions that follow.

A neuron's unique structure enables it to conduct electrical signals called nerve impulses. Neurons communicate by transmitting nerve impulses to body tissues and organs, including muscles, glands, and other neurons. Neurons vary greatly in form. **Dendrites,** which extend from the cell body of the neuron, are the "antennae" of the neuron. Dendrites receive information from other cells. The neuron's cell body collects information from the dendrites, relays this information to other parts of the neuron, and maintains the general functioning of the neuron. An **axon** is a long membrane-covered extension of the cytoplasm that conducts nerve impulses. The ends of an axon are called axon terminals. When a neuron communicates with other cells, it does so at its axon terminals.

Nervous tissue consists mostly of neurons and their supporting cells. Bundles of axons are called **nerves**. The arrangement of axons in a nerve is similar to a telephone cable with many different communication channels, each carried by a separate wire. Nerves appear as fine, white threads when viewed with the unaided eye.

SKILL: READING EFFECTIVELY

Read each question, and write your answer in the space provided.

1. How do neurons communicate?

____

____

2. What is the function of a neuron's dendrites?

____

____

3. What is the function of a neuron's cell body?

____

____

____

4. What is an axon?

____

____

5. How are axons and nerves related?

____

____

In the space provided, write the letter of the term or phrase that best completes the statement.

______ **6.** A neuron is able to conduct electrical signals, or nerve impulses, due to its unique

a. size.
b. location.
c. structure.
d. Both (a) and (b)

Name ______________________ Class ____________ Date ___________

Skills Worksheet

Active Reading

Section: Structures of the Nervous System

Read the passage below. Then answer the questions that follow.

The brain consists of three major parts—the cerebrum, the cerebellum, and the brain stem.

Cerebrum: The **cerebrum** is the largest part of the brain. The capacity for learning, memory, perception, and intellectual functioning resides in the cerebrum. The cerebrum has a folded outer layer with many bumps and grooves. A long, deep groove down the center of the brain divides the cerebrum into right and left halves, or hemispheres. The cerebral hemispheres communicate through a connecting band of axons called the corpus callosum. In general, the left cerebral hemisphere receives sensations from and controls the movements of the right side of the body. The right cerebral hemisphere receives sensations from and controls the movements of the left side of the body.

Most sensory and motor processing occurs in the cerebral cortex, the folded, thin (2–4 mm) outer layer of the cerebrum. The cerebral cortex contains about 10 percent of the brain's neurons. The folded outer surface of the cerebrum is the cerebral cortex, which has a large surface area. The cerebral cortex is primarily involved with the functioning of sensory systems.

Cerebellum: The **cerebellum,** which is located at the posterior base of the brain, regulates balance, posture, and movement. The cerebellum smoothes and coordinates ongoing movements, such as walking, by timing the contraction of skeletal muscles. The cerebellum integrates and responds to information about body position from the cerebrum and the spinal cord to control balance and posture.

Brain stem: At the base of the brain is the stalklike **brain stem.** The brain stem is a collection of structures leading down to the spinal cord and connecting the cerebral hemispheres with the cerebellum. The lower brain stem consists of the midbrain, the pons, and the medulla oblongata. These structures relay information throughout the central nervous system and play an important role in homeostasis by regulating vital functions such as heart rate, breathing rate, body temperature, and sleep.

SKILL: READING EFFECTIVELY

In the space provided, write the letter of the part of the brain that best matches the description.

a. cerebrum
b. cerebellum
c. brain stem

______ **1.** regulates balance

______ **2.** leads to the spinal cord

______ **3.** largest part of the brain

______ **4.** divided into two hemispheres

______ **5.** pons is located here

______ **6.** contains the corpus callosum

______ **7.** responds to information about body position

______ **8.** capacity for learning resides here

______ **9.** breathing rate is regulated here

______ **10.** located at the base of the brain

______ **11.** has a folded layer with many bumps and grooves

______ **12.** regulates posture

______ **13.** left side controls right side of the body

______ **14.** medulla oblongata located here

______ **15.** times contractions of the skeletal muscles

______ **16.** cerebral cortex is located here

______ **17.** located at the posterior base of the brain

______ **18.** heart rate is regulated here

______ **19.** capacity for intellectual function resides here

An analogy is a comparison. In the space provided, write the letter of the term or phrase that best completes the analogy.

______ **20.** Cerebellum is to movement as cerebrum is to
a. breathing.
b. walking.
c. sleeping.
d. perception.

Name ______________________ Class ______________ Date ____________

Skills Worksheet

Active Reading

Section: Sensory Systems

Read the passage below. Then answer the questions that follow.

Your ears convert the energy in sound waves to electrical signals that can be interpreted by your brain. Sound waves enter through the ear canal and strike the tympanic membrane, or eardrum, causing the tympanic membrane to vibrate. Behind the eardrum, three small bones of the middle ear—the hammer, anvil, and stirrup—transfer the vibrations to a fluid-filled chamber within the inner ear. This chamber, called the **cochlea,** is coiled like a snail's shell, and it contains mechanoreceptors called hair cells. Hair cells rest on a membrane that vibrates when waves enter the cochlea. Waves of different frequencies cause different parts of the membrane to vibrate and thus stimulate different hair cells. When hair cells are stimulated, they generate nerve impulses in the auditory nerve. The impulses travel to the brain stem through the auditory nerve. The thalamus then relays the information to the temporal lobe of the cerebral cortex, where the auditory information is processed.

The ears not only enable you to hear but also help you maintain equilibrium. The **semicircular canals** are fluid-filled chambers in the inner ear that contain hair cells. Clusters of these hair cells respond to changes in head position with respect to gravity. When your head moves, the hair cells are stimulated according to the magnitude and direction of the fluid's movement and send electrical signals to the brain. Signals generated by the hair cells enable the brain to determine the orientation and position of the head.

SKILL: READING EFFECTIVELY

Read each question, and write your answer in the space provided.

1. According to the passage, what are two functions of the ear?

____

____

2. How does energy change form in the ear?

____

____

3. What causes the tympanic membrane to vibrate?

____

____

____

4. What is the cochlea?

____

____

5. How do vibrations reach the cochlea?

____

____

____

6. What occurs when hair cells are stimulated?

____

____

7. What are semicircular canals?

____

____

8. How do the semicircular canals help you maintain balance?

____

____

____

____

In the space provided, write the letter of the term or phrase that best completes the statement.

______ **9.** The semicircular canals and cochlea are alike in that both

a. are located in the inner ear.
b. are fluid-filled chambers.
c. convert sound energy to electrical signals.
d. Both (a) and (b)

Name ______________________________ Class ______________ Date ____________

Skills Worksheet

Active Reading

Section: Drugs and the Nervous System

Read the passage below. Then answer the questions that follow.

Alcohol, found in wine, beer, and liquor, is a depressant that produces a sense of well-being when taken in small amounts. A **depressant** is a drug that generally decreases the activity of the central nervous system. As more alcohol is consumed, reaction time increases, and coordination, judgment, and speech become impaired. This produces a state of intoxication known as being "drunk."

Alcohol is absorbed into the blood through the stomach and small intestine. Alcohol affects neurons throughout the nervous system, changing the shape of receptor proteins. Such widespread changes in receptor proteins have various effects on normal brain functioning.

SKILL: READING EFFECTIVELY

Read each question, and write your answer in the space provided.

1. What is a depressant?

____

____

2. What occurs as the amount of alcohol taken into the body increases?

____

____

3. How does alcohol enter the bloodstream?

____

____

In the space provided, write the letter of the term or phrase that best completes the statement.

______ **4.** Alcohol is found in all of the following EXCEPT

a. cola.
b. wine.
c. liquor.
d. beer.

Name ______________________________ Class ______________ Date ____________

Skills Worksheet

Vocabulary Review

In the space provided, write the letter of the description that best matches the term or phrase.

______ **1.** neuron

______ **2.** dendrite

______ **3.** axon

______ **4.** nerve

______ **5.** membrane potential

______ **6.** resting potential

______ **7.** action potential

______ **8.** synapse

______ **9.** neurotransmitter

a. the difference in electrical charge across a cell membrane

b. part of a neuron that conducts nerve impulses

c. the membrane potential of a neuron at rest

d. nerve cell; transmits information throughout the body

e. bundle of axons

f. part of a neuron that receives information from other neurons

g. a junction at which a neuron meets another cell

h. a signal molecule that transmits nerve impulses across synapses

i. nerve impulse

Write the correct term from the list below in the space next to its definition.

brain	hypothalamus	reflex
brain stem	interneurons	sensory neuron
central nervous system	motor neuron	spinal cord
cerebellum	peripheral nervous system	thalamus
cerebrum		

______________________ **10.** sends commands from the central nervous system to muscles and other organs

______________________ **11.** site of capacity for learning, memory, perception, and intellectual function

______________________ **12.** consists of the brain and spinal cord

______________________ **13.** relays sensory information to the cerebral cortex

______________________ **14.** dense cable of nervous tissue that runs through the vertebral column

______________________ **15.** contains sensory neurons and motor neurons

______________________ **16.** carries information from sense organs to the central nervous system

______________________ **17.** the body's main processing center

______________________ **18.** helps regulate breathing, heart rate, and endocrine functions

______________________ **19.** link neurons to each other

______________________ **20.** collection of structures leading down to the spinal cord

______________________ **21.** regulates balance, posture, and movement

______________________ **22.** a sudden, involuntary contraction of muscles in response to a stimulus

In the space provided, write the letter of the description that best matches the term or phrase.

______ **23.** sensory receptor

______ **24.** retina

______ **25.** rod

______ **26.** cone

______ **27.** optic nerve

______ **28.** cochlea

______ **29.** semicircular canal

a. the lining of photoreceptors and neurons in the eye

b. aids in hearing

c. type of photoreceptor that responds best to dim light

d. runs from the back of each eye to the brain

e. helps maintain equilibrium

f. a specialized neuron that detects sensory stimuli

g. type of photoreceptor that enables color vision

Complete each statement by writing the correct term or phrase in the space provided.

30. The need for increasing amounts of a drug to achieve the desired sensation is called ______________________ .

31. A drug that generally decreases the activity of the central nervous system is called a(n) ______________________ .

32. A drug that generally increases the activity of the central nervous system is called a(n) ______________________ .

33. Drugs that alter the functioning of the central nervous system are known as

______________________ ______________________ .

34. A set of emotional and physical symptoms caused by removing a drug from the body of a drug addict is ______________________ .

35. A physiological response caused by use of a drug that alters the normal functioning of neurons and synapses is ______________________ .

Name ______________________ Class ______________ Date ______________

Skills Worksheet

Science Skills

Interpreting Data

Neurons are classified into three groups based on the diameter of their axons—A fibers, B fibers, and C fibers. A fibers, the largest, have axon diameters of 5–20 µm and are myelinated. B fibers have axon diameters of 2–3 µm and are also myelinated. C fibers have the smallest axon diameters of only 0.5–1.5 µm. C fibers are not myelinated. The data table below shows the results of experiments conducted on the nerve fibers of vertebrate animals. Use this information and the results in the table below to answer questions 1–3.

Data Table

Fiber Class	Fiber Diameter (µm)	Conduction Speed (m/s)
A_1	18.5	42
A_2	14.0	25
A_3	11.0	17
B	2.0	10
C	0.5	0.5

Read each question, and write your answer in the space provided.

1. What conclusions can be reached about the relationship between the diameter of an axon and the speed with which it conducts impulses?

____

____

2. Hypothesize which types of impulses are carried by A fibers. Explain your answer.

____

____

____

____

____

3. Invertebrates do not have myelinated axons. However, some invertebrates, such as squid, have giant nerve fibers with axons up to 1 mm in diameter. Other large diameter axons are found in crayfish and earthworms. Explain why these organisms have large axons. Predict which types of impulses might be carried by these large axons.

____

____

____

____

Interpreting Scientific Illustrations

The figure below shows a synapse in which an addictive drug is present. In the space provided, write the name of the following structures: reuptake receptor, neurotransmitter molecule, axon terminal, receptor protein, and drug molecule.

________________________ **4.**

________________________ **5.**

________________________ **6.**

________________________ **7.**

________________________ **8.**

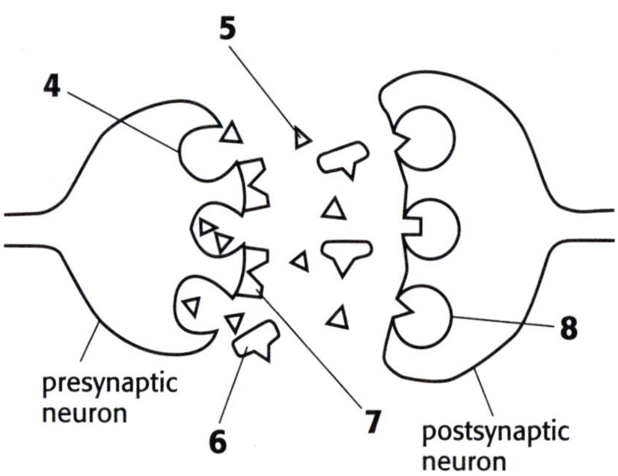

9. How does the drug molecule interfere with the events taking place at the synapse? How does this result in the postsynaptic neuron being overstimulated?

____

____

____

____

____

Name ______________________ Class ______________ Date ____________

Skills Worksheet

Concept Mapping

Using the terms and phrases provided below, complete the concept map showing the structures of the nervous system.

central nervous system	parasympathetic division	spinal cord
interneurons	peripheral nervous system	sympathetic division
motor neurons	somatic nervous system	synapses
neurons		

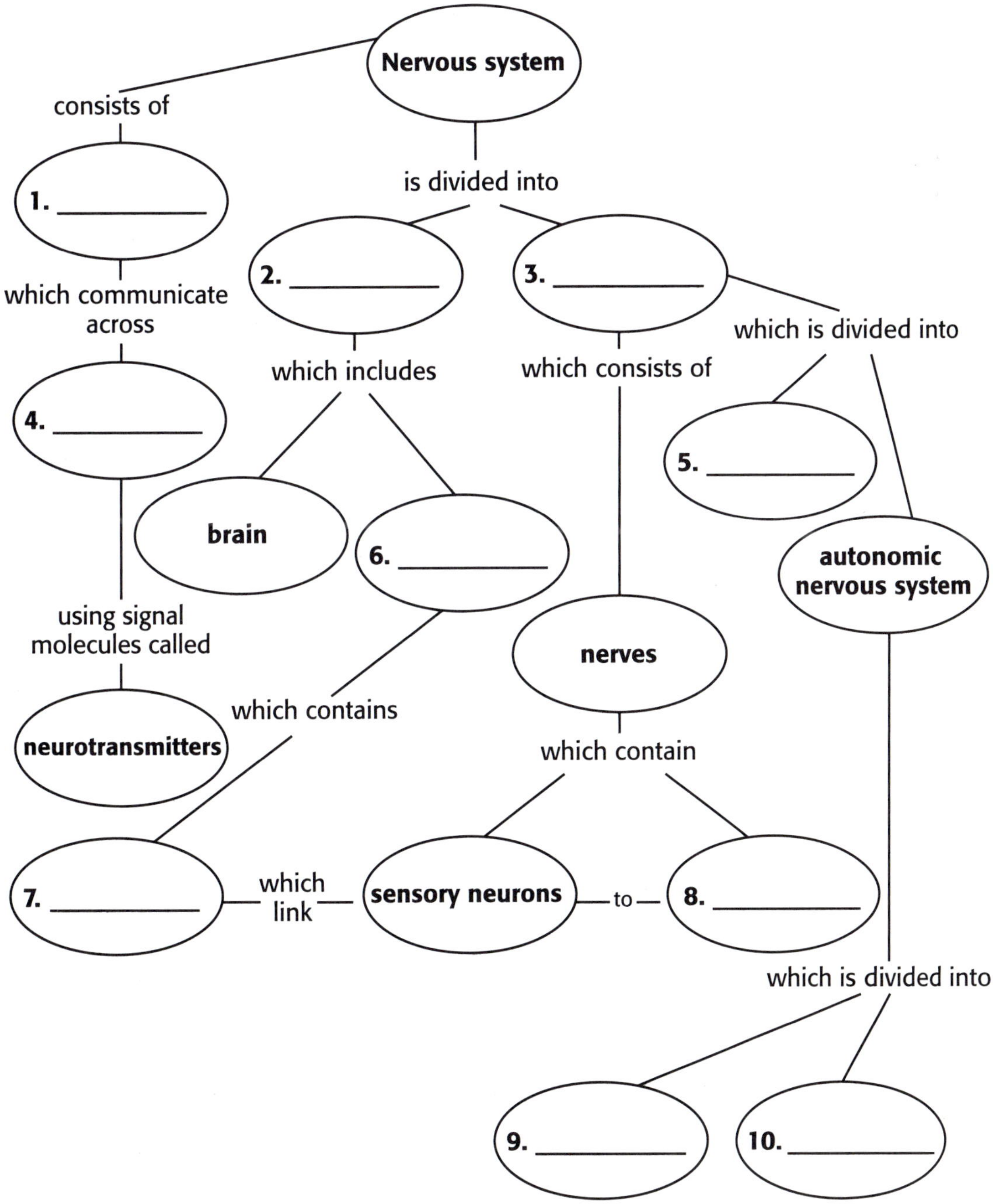

Name ______________________ Class ______________ Date ____________

Skills Worksheet

Critical Thinking

Look-Alikes

In the space provided, write the letter of the term or phrase that best describes how each numbered item looks.

______ **1.** neuron cell body with dendrites

______ **2.** gray matter in the spinal cord

______ **3.** cerebrum

______ **4.** cross-section of the cerebellum

______ **5.** taste bud

______ **6.** cochlea

a. wrinkled boxing glove

b. snail shell

c. small head of cauliflower

d. butterfly wings

e. uprooted tree

f. flower bud

Work-Alikes

In the space provided, write the letter of the term or phrase that best describes how each numbered item functions.

______ **7.** voltage-gated ion channels

______ **8.** action potential

______ **9.** neurotransmitter molecules

______ **10.** human eye

______ **11.** spinal cord

______ **12.** enkephalin

a. two-way highway

b. flame burning down a fuse

c. flood gates

d. narcotic

e. camera

f. ferry boats

Cause and Effect

In the space provided, write the letter of the term or phrase that best matches each cause or effect given below.

Cause	Effect
13. presence of myelin	______________
14. difference in charge	______________
15. ______________	it opens and ions enter the neuron
16. ______________	skeletal muscle movement
17. stimulation of autonomic nervous system	______________

a. neurotransmitter binds to chemical-gated channel

b. stimulation of somatic nervous system

c. smooth muscle regulation

d. membrane potential

e. nerve impulses move faster

Linkages

In the spaces provided, write the letters of the two terms or phrases that are linked together by the term or phrase in the middle. The choices can be placed in any order.

18. ______ sodium ions flow into neuron ______

19. ______ limbic system ______

20. ______ interneurons ______

21. ______ cocaine blocks ______

22. ______ nicotine binds to acetylcholine sites of nerve cells ______

23. ______ smoking-related illnesses ______

a. cerebral cortex
b. tobacco use
c. motor neurons
d. voltage-gated sodium channels open
e. dopamine stays in synapse, producing feelings of well-being
f. 400,000 deaths per year
g. smoking
h. change at these sites; addiction results
i. thalamus and hypothalamus
j. action potential
k. sensory neurons
l. cocaine use

Analogies

An analogy is a relationship between two pairs of terms or phrases written as a : b :: c : d. The symbol : is read as "is to," and the symbol :: is read as "as." In the space provided, write the letter of the pair of terms or phrases that best completes the analogy shown.

______**24.** myelin sheath : axon ::
a. cell body : dendrites
b. axon : dendrite
c. nucleus : cell body
d. cell body : axon

______**25.** central nervous system : brain and spinal cord ::
a. brain : central nervous system
b. motor nerves : nerves that receive messages from muscles and glands
c. spinal cord : cerebrum and cerebral cortex
d. peripheral nervous system : sensory nerves and motor nerves

______**26.** hair cells : semicircular canals ::
a. hair cells : outer ear
b. anvil : outer ear
c. hair cells : cochlea
d. hammer : cochlea

______**27.** drug use : addiction ::
a. drug removal : withdrawal
b. tolerance : withdrawal
c. withdrawal : addition of drug to body
d. withdrawal : tolerance

Name ______________________ Class ______________ Date ____________

Skills Worksheet

Test Prep Pretest

In the space provided, write the letter of the term or phrase that best completes each statement or best answers each question.

______ **1.** During a knee-jerk reflex, the nerve impulse is received by the
a. brain.
b. spinal cord.
c. spinal cord and then the brain.
d. thalamus.

Questions 2–4 refer to the figure below, which shows the structure of a typical neuron.

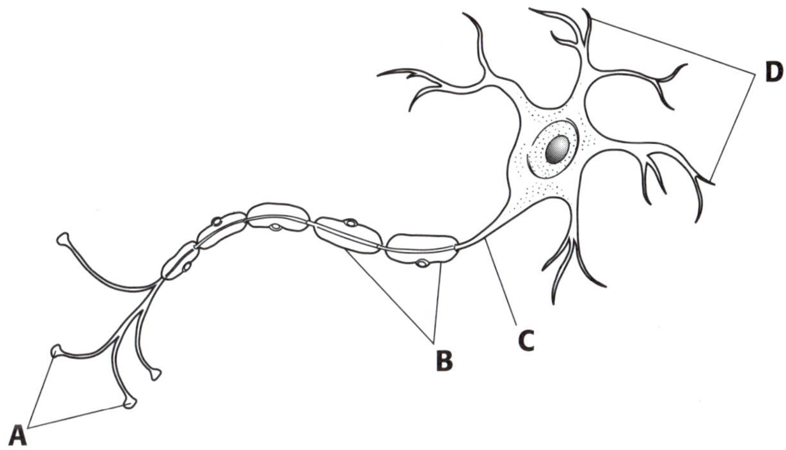

______ **2.** What occurs when an action potential reaches the structures labeled *A*?
a. Neurotransmitters are released.
b. Myelin sheaths are activated.
c. Cell bodies receive messages.
d. None of the above

______ **3.** The structures labeled *B* are
a. axon terminals.
b. nodes of Ranvier.
c. dendrites.
d. myelin sheaths.

______ **4.** The structures labeled *D* are
a. dendrites.
b. axons.
c. axon terminals.
d. nodes of Ranvier.

______ **5.** Light entering the eye stimulates
a. hair cells in the retina.
b. the optic nerve.
c. rods and cones in the retina.
d. mechanoreceptors.

______ **6.** When cocaine interferes with reuptake receptors on a presynaptic neuron, the
a. postsynaptic cell is overstimulated.
b. number of neurotransmitter receptors decreases.
c. excess neurotransmitters remain in the synaptic cleft.
d. All of the above

______ **7.** Drug use that alters normal functioning of neurons and synapses results in
a. addiction.
b. withdrawal.
c. tolerance.
d. None of the above

Questions 8–11 refer to the figure at right.

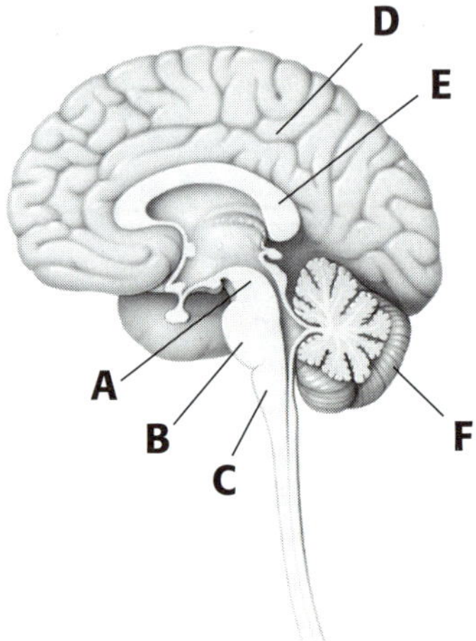

______ **8.** The structures labeled *A*, *B*, and *C* regulate
a. heart rate and sleep.
b. body temperature.
c. breathing rate.
d. All of the above

______ **9.** The structure labeled *D* is involved in
a. balance and posture.
b. maintaining homeostasis.
c. learning, memory, and perception.
d. spinal reflexes.

______ **10.** The structure labeled *E* is the
a. thalamus.
b. corpus callosum.
c. brain stem.
d. cerebrum.

______ **11.** The structure labeled *F* is the
a. thalamus.
b. hypothalamus.
c. cerebellum.
d. cerebral cortex.

Complete each statement by writing the correct term or phrase in the space provided.

12. The ______________________ ______________________ of a neuron is negative because there are more positively charged ions outside the cell than inside the cell.

13. A(n) ______________________ ______________________ is a local reversal of polarity inside a neuron.

14. During synaptic transmission, a presynaptic neuron releases a(n) ______________________ into the synaptic ______________________ .

15. At a synapse, a neurotransmitter may ______________________ or ______________________ the activity of the postsynaptic cell it binds to.

16. After a nerve impulse has passed, ______________________ ions flow out of the axon, and the membrane potential becomes ______________________ again.

17. In the spinal cord, cell bodies of neurons make up the ______________________ matter, while axons make up the ______________________ matter.

18. The ______________________ nervous system contains neurons that connect the brain and the spinal cord to the rest of the body.

19. The coiled inner ear structure that converts sound waves to nerve impulses is called the ______________________ .

20. Pain receptors are located on all tissues and organs except the ______________________ .

21. Auditory information is processed in the ______________________ ______________________ of the brain.

22. Drugs that alter the functioning of the central nervous system and are often addictive are called ______________________ ______________________ .

23. A(n) ______________________ is a substance that decreases the activity of the central nervous system.

Read each question, and write your answer in the space provided.

24. Distinguish between the somatic nervous system and the autonomic nervous system.

____

____

____

____

____

____

25. List the four basic types of chemicals that taste buds can detect.

____

____

Name ______________________________ Class ______________ Date ____________

Assessment

Quiz

Section: Neurons and Nerve Impulses

The figure below shows a myelinated neuron. In the space provided, write the letter of the name of the following structures:

a. myelin sheaths

b. axon terminals

c. dendrites

d. nucleus

e. cell body

f. nodes of Ranvier

g. axon

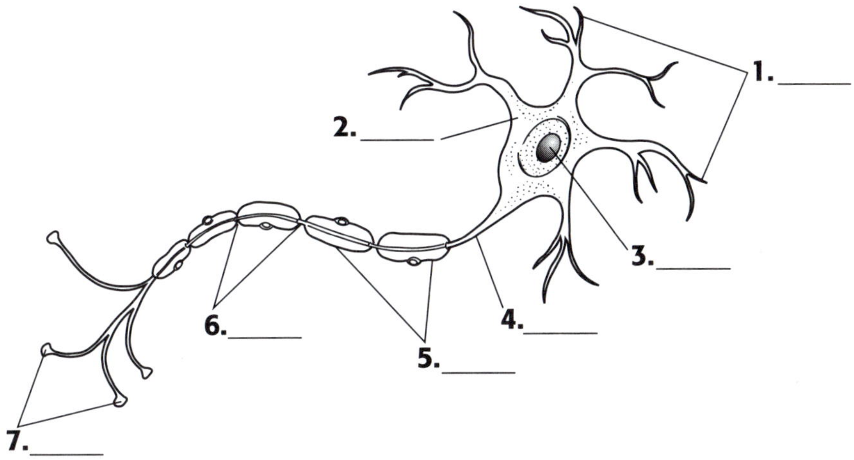

In the space provided, write the letter of the term or phrase that best completes each statement or best answers each question.

______ **8.** The conduction of nerve impulses is faster in myelinated axons because

a. neurotransmitter molecules are released in greater amounts.

b. the myelin sheath covers the entire axon.

c. nerve impulses "jump" from node to node as they move down the axon.

d. the membrane potential does not change.

______ **9.** The resting potential of a neuron is usually which of the following?

a. positive

b. negative

c. neutral

d. Either (a) or (b)

______ **10.** Neurotransmitter molecules are removed from a synapse by

a. being broken down by enzymes.

b. being reabsorbed by the presynaptic neuron.

c. being absorbed by muscle tissue.

d. Both (a) and (b)

Name ______________________ Class ______________ Date ____________

Assessment

Quiz

Section: Structures of the Nervous System

In the space provided, write the letter of the term or phrase that best completes each statement or best answers each question.

_______ **1.** The two main divisions of the nervous system are
- **a.** the autonomic nervous system and the somatic nervous system.
- **b.** the sympathetic division and the parasympathetic division.
- **c.** the central nervous system and the peripheral nervous system.
- **d.** the sensory division and the motor division.

_______ **2.** Neurons that link other neurons together are called
- **a.** communicating neurons.
- **b.** interneurons.
- **c.** sensory neurons.
- **d.** motor neurons.

_______ **3.** Which of the following divisions of the autonomic nervous system causes heart rate to increase, pupils to dilate, and blood vessels to dilate?
- **a.** parasympathetic division
- **b.** somatic division
- **c.** sensory division
- **d.** sympathetic division

Questions 4–10 refer to the figure below, which shows a cross-section of the brain and spinal cord. In the space provided, write the letter of the name of the part of the brain and spinal cord.

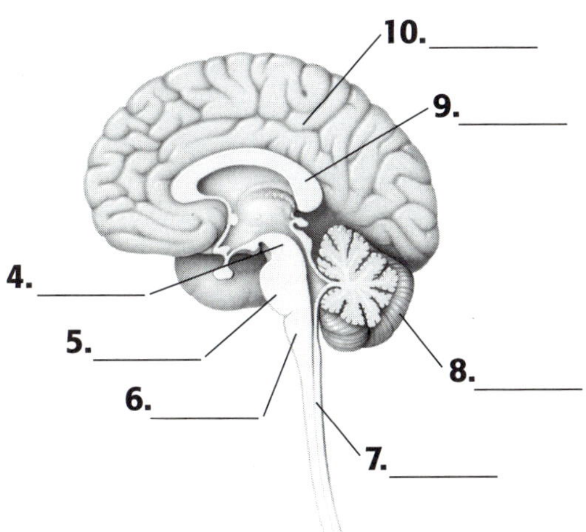

- **a.** cerebrum
- **b.** spinal cord
- **c.** midbrain
- **d.** corpus callosum
- **e.** pons
- **f.** cerebellum
- **g.** medulla oblongata

Name ______________________ Class ______________ Date ____________

Assessment

Quiz

Section: Sensory Systems

In the space provided, write the letter of the term or phrase that best completes each statement or best answers each question.

______ **1.** Sensory information from the olfactory receptors is processed in which lobe of the brain?

a. parietal lobe
b. frontal lobe
c. occipital lobe
d. temporal lobe

______ **2.** The area of the eye that contains photoreceptors and neurons is the

a. cornea.
b. lens.
c. retina.
d. iris.

______ **3.** The cochlea is located

a. in the outer ear.
b. in the taste buds.
c. in the inner ear.
d. on the tongue's surface.

______ **4.** Semicircular canals contain which of the following structures, which are stimulated by changes in the position of the head with respect to gravity?

a. taste buds
b. hair cells
c. the hammer, the anvil, and the stirrup
d. rods and cones

______ **5.** Sensory receptors are most concentrated in which of the following areas of the body?

a. eyes, ears, and skin
b. muscle tissue of the arms and legs
c. nose and mouth
d. Both (a) and (c)

In the space provided, write the letter of the description that best matches the term or phrase.

______ **6.** detect movement, pressure, and tension

______ **7.** located in all tissues except the brain

______ **8.** are stimulated by light

______ **9.** located in the tongue and nose

______ **10.** detect changes in temperature

a. thermoreceptors
b. pain receptors
c. mechanoreceptors
d. photoreceptors
e. chemoreceptors

Name ______________________ Class ______________ Date ____________

Assessment

Quiz

Section: Drugs and the Nervous System

In the space provided, write the letter of the term or phrase that best completes each statement or best answers each question.

______ **1.** Tars, produced from burning tobacco, do all of the following EXCEPT
a. accumulate in the lungs and paralyze cilia.
b. blacken lung tissue and decrease breathing ability.
c. irritate mucous membranes in the mouth, nose, and throat.
d. cause addiction to cigarettes.

______ **2.** Neurotransmitters called enkephalins
a. prevent pain signals from reaching the brain.
b. cause pain signals to reach the brain.
c. stimulate pain receptors to send impulses.
d. None of the above

______ **3.** Cirrhosis, a condition in which liver cells are replaced with scar tissue, can develop as a result of
a. alcoholism.
b. using stimulants.
c. malnutrition.
d. smoking cigarettes.

______ **4.** Examples of psychoactive drugs include which of the following?
a. caffeine
b. alcohol
c. cocaine
d. All of the above

In the space provided, write the letter of the description that best matches the term or phrase.

______ **5.** depressants

______ **6.** stimulants

______ **7.** inhalants

______ **8.** hallucinogens

______ **9.** THC

______ **10.** narcotics

a. examples include heroin, morphine, and codeine
b. psychoactive effects include sensory distortion, hallucinations, and numbness
c. psychoactive effects include disorientation, confusion, and memory loss
d. risks associated with use include lung damage and loss of motivation
e. examples include tranquilizers and alcohol
f. psychoactive effects include increased activity of the central nervous system

Name ______________________________ Class ______________ Date ____________

Assessment

Chapter Test

Nervous System

In the space provided, write the letter of the term or phrase that best completes each statement or best answers each question.

______ **1.** The brain and spinal cord make up the
- **a.** central nervous system.
- **b.** peripheral nervous system.
- **c.** autonomic nervous system.
- **d.** None of the above

______ **2.** In times of stress, the division of the autonomic nervous system that dominates is the
- **a.** sympathetic nervous system.
- **b.** parasympathetic nervous system.
- **c.** motor division.
- **d.** sensory division.

______ **3.** Pressure is detected by
- **a.** thermoreceptors.
- **b.** photoreceptors.
- **c.** mechanoreceptors.
- **d.** chemoreceptors.

______ **4.** Alcohol consumption can
- **a.** alter neurons throughout the nervous system.
- **b.** affect normal brain function.
- **c.** cause abnormalities in the circulatory system.
- **d.** All of the above

______ **5.** The peripheral nervous system connects the body to the
- **a.** upper brain stem.
- **b.** hypothalamus.
- **c.** brain and spinal cord.
- **d.** autonomic nervous system.

______ **6.** Myelin sheaths
- **a.** are found on all neurons.
- **b.** increase the speed of nerve impulses.
- **c.** decrease the speed of nerve impulses.
- **d.** All of the above

______ **7.** Neurons
- **a.** transmit information throughout the body.
- **b.** enable functions such as movement, perception, emotion, and learning.
- **c.** carry ions throughout the body.
- **d.** Both (a) and (b)

______ **8.** Which part of the brain smoothes and coordinates movements such as walking?
- **a.** cerebrum
- **b.** cerebellum
- **c.** brain stem
- **d.** hypothalamus

______ **9.** Addiction to a drug occurs because
a. increasing amounts of the drug are needed to achieve the desired sensation.
b. the drug is removed from the body.
c. the drug alters the normal functioning of neurons and synapses so that they cannot function normally unless the drug is present.
d. drugs stimulate the central nervous system.

______ **10.** Each spinal nerve has
a. a dorsal root and a ventral root.
b. sensory neurons.
c. motor neurons.
d. All of the above

______ **11.** The psychoactive effects of hallucinogens include
a. increased activity of the central nervous system.
b. sensory distortion, anxiety, and hallucinations.
c. decreased activity of the central nervous system.
d. short-term memory loss and impaired judgment.

______ **12.** Visual information is processed by which lobe of the cerebrum?
a. occipital
b. parietal
c. temporal
d. frontal

______ **13.** Olfactory receptors and taste cells are both
a. mechanoreceptors.
b. located in the roof of the nasal passage.
c. chemoreceptors.
d. found in taste buds.

______ **14.** A spinal reflex is
a. a self-protective motor response.
b. rapid because it usually does not involve the brain.
c. involuntary.
d. All of the above

______ **15.** semicircular canals : balance ::
a. retina : hearing
b. cochlea : hearing
c. cochlea : sight
d. taste cells : smell

______ **16.** The axons that carry information from the retina to the brain are called
a. the optic nerve
b. the auditory nerve
c. rods and cones
d. photoreceptors

In the space provided, write the letter of the description that best matches the term or phrase.

______ **17.** synapse

______ **18.** resting potential

______ **19.** neurotransmitters

______ **20.** action potential

a. the sodium-potassium pump helps maintain the inside of the cell as negatively charged compared to the outside of the cell

b. occurs when a stimulus causes a positive change in the membrane potential near an axon

c. nerve impulses have to cross this junction, which occurs between neurons

d. molecules that allow an impulse to cross the synapse

Name ______________________ Class ____________ Date ___________

Assessment

Chapter Test

Nervous System

In the space provided, write the letter of the term or phrase that best completes each statement or best answers each question.

______ **1.** Which of the following is the control center of the body?
- **a.** the peripheral nervous system
- **b.** the central nervous system
- **c.** the autonomic nervous system
- **d.** the somatic nervous system

______ **2.** The photoreceptors of the retina
- **a.** convert light energy into electrical signals.
- **b.** convert sound waves into electrical signals.
- **c.** convert light energy into chemical signals.
- **d.** convert chemical signals into electrical signals.

______ **3.** Which of the following connects the cerebral hemispheres with the cerebellum?
- **a.** thalamus
- **b.** hypothalamus
- **c.** brain stem
- **d.** spinal cord

______ **4.** The function of the semicircular canals is to
- **a.** help maintain equilibrium.
- **b.** convert sound waves to electrical signals.
- **c.** enable the brain to determine the orientation and position of the head.
- **d.** Both (a) and (c)

______ **5.** The myelin sheath
- **a.** transmits impulses.
- **b.** insulates synapses.
- **c.** insulates axons.
- **d.** None of the above

______ **6.** The two principal components of the peripheral nervous system are the
- **a.** somatic and the autonomic nervous systems.
- **b.** sensory and motor divisions.
- **c.** autonomic and central nervous systems.
- **d.** sympathetic and the parasympathetic divisions.

______ **7.** Motor neurons that stimulate skeletal muscles are part of the
- **a.** somatic nervous system.
- **b.** autonomic nervous system.
- **c.** parasympathetic nervous system.
- **d.** corpus callosum.

______ **8.** Nicotine mimics the action of the neurotransmitter
- **a.** dopamine.
- **b.** acetylcholine.
- **c.** glutamate.
- **d.** None of the above

______ **9.** Narcotics
a. are extremely addictive.
b. mimic natural painkillers called enkephalins.
c. affect the limbic system, producing a feeling of well-being.
d. All of the above

______ **10.** Nitrous oxide, ether, paint thinner, and glue belong to a class of psychoactive drugs called
a. stimulants.
b. inhalants.
c. depressants.
d. hallucinogens.

______ **11.** When a neuron is at rest,
a. sodium ions are more concentrated outside the cell.
b. potassium ions are more concentrated inside the cell.
c. the inside of the cell is negatively charged.
d. All of the above

Questions 12 and 13 refer to the figure below.

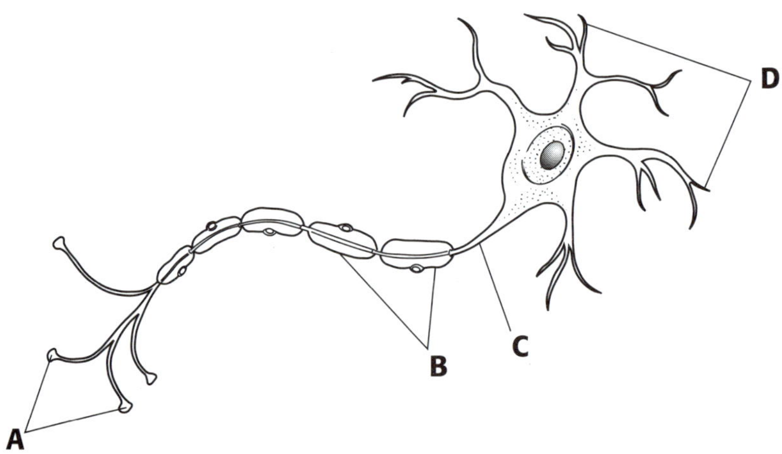

______ **12.** The function of the structures labeled *D* is which of the following?
a. receive information from other cells
b. conduct nerve impulses
c. maintain the general functioning of the neuron
d. make impulses move faster

______ **13.** An action potential is conducted down the axon, which is labeled
a. *A*.
b. *B*.
c. *C*.
d. *D*.

______ **14.** During a nerve impulse,
a. sodium ions first rush out of the cell.
b. there is a reversal of polarity in the axon.
c. the membrane potential of the cell does not change.
d. potassium ions are pumped into the axon.

Complete each statement by writing the correct term or phrase in the space provided.

15. A(n) ______________________ ______________________ is a cluster of 50–100 taste cells.

16. The gray matter in the spinal cord contains the ______________________ of neurons, whereas the white matter contains the ______________________ of neurons.

17. During an action potential, ______________________ channels open, allowing ______________________ ions to move into the axon.

18. Sensory receptors that respond to physical stimuli that cause distortion or bending of tissue are ______________________ .

19. Chemoreceptors that detect odors and are located in the roof of the nasal passage are called ______________________ receptors.

20. Sensory information about hearing is processed in the ______________________ lobe of the brain.

Read each question, and write your answer in the space provided.

21. How is a signal from a presynaptic neuron transmitted to a postsynaptic cell?

____

____

____

____

____

____

22. Describe the function of the cerebrum.

____

____

____

____

23. Explain how drug use can lead to drug addiction.

____

____

____

____

____

24. Describe how the ear converts sound waves into electrical impulses that the brain can process.

____

____

____

____

____

____

25. Describe the action of cocaine at a synapse.

____

____

____

____

____

Name ______________________ Class ______________ Date ______________

Data Lab **DATASHEET FOR IN-TEXT LAB**

Analyzing Changes During a Nerve Impulse

Background

The graph below illustrates changes that occur in the membrane potential of a neuron during an action potential. Use the graph to answer the following questions.

Action Potential

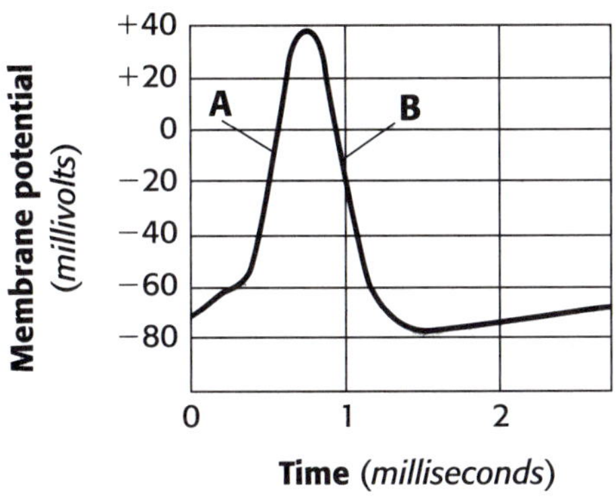

Analysis

1. **Determine** about how long an action potential lasts.

____

2. **State** whether voltage-gated sodium channels are open or closed at point *A*.

____

3. **State** whether voltage-gated potassium channels are open or closed at point *B*.

____

4. **Critical Thinking**
Recognizing Relationships What causes the membrane potential to become less negative at point *A*?

____

____

5. **Critical Thinking**
Recognizing Relationships What causes the membrane potential to become more negative at point *B*?

____

____

Name ______________________ Class ______________ Date ____________

Quick Lab

DATASHEET FOR IN-TEXT LAB

Demonstrating Your Blind Spot

The blind spot in your visual field corresponds to the site where the optic nerve exits the back of the eye. There are no photoreceptors at the site where the optic nerve exits. Use the procedure below to demonstrate your blind spot.

MATERIALS

- unlined 3 × 5 index card
- pencil

Procedure

1. On the index card, draw an *X* about 1 in. from the left side of the card. Draw an *O* about the same size 3 in. to the right of the *X*.
2. Hold your index card in front of you at arm's length. Close your right eye and stare at the *O* with your left eye. Slowly move the card toward you while continuing to stare at the *O* until the *X* disappears from view.

Analysis

1. **Name** the two kinds of photoreceptors found in the retina.

____

____

2. **Propose** why you cannot see images that fall on the site where the optic nerve exits the eye.

____

____

3. **Critical Thinking**
Relating Concepts What is the relationship between the structure of the retina and the disappearance of the *X* on the index card?

____

____

____

Name ______________________________ Class ______________ Date ____________

Exploration Lab

DATASHEET FOR IN-TEXT LAB

Calculating Reaction Times

SKILLS

- Measuring
- Calculating

OBJECTIVES

- **Determine** human reaction times.
- **Design** an experiment that measures changes in reaction times.

MATERIALS

- meterstick

Before You Begin

When you want to move your hand, your brain must send a message all the way to the muscles in your arms. How long does that take? In this exercise, you will work with a partner to see how quickly you can react. In this lab, you will investigate reaction times and design an experiment to investigate influences on reaction times.

1. You will be using the data table provided to record your data.
2. Write a hypothesis about an influence on reaction times. (For example: "People who have eaten breakfast have faster reaction times than people who have not eaten breakfast.")

____

____

____

Procedure

PART A: CALCULATING REACTION TIMES

1. Sit in a chair and have a partner stand facing you while holding a meterstick in a vertical position.
2. Hold your thumb about 3 cm from your fingers near the bottom end of the stick. The meterstick should be positioned to fall between your thumb and fingers.
3. Tell your partner to let go of the meterstick without warning.
4. When your partner releases the meterstick, catch the stick by pressing your thumb and fingers together. Your partner should be ready to catch the top of the meterstick if it begins to tip over.

5. Record the number of centimeters the stick dropped before you caught it. The distance that the meterstick falls before you catch it can be used to evaluate your reaction time.

6. Repeat the procedure several times, and calculate the average number of centimeters.

7. Try this procedure with your other hand.

8. Close your eyes and have your partner say "now," when the stick is released.

9. Exchange places with your partner, and repeat the procedure.

PART B: DESIGNING YOUR OWN EXPERIMENT

10. Work with the members of your lab group to explore one of the hypotheses written in the **Before You Begin** section of this lab.

> **You Choose**
> As you design your experiment, decide the following:
> **a.** what hypothesis you will explore
> **c.** how you will test the hypothesis
> **d.** what the controls will be
> **e.** how many trials to perform
> **f.** what data to record in your data table

Data Table		
Hand: trial number	**Subject 1 reaction time (s)**	**Subject 2 reaction time (s)**
Left: 1		
Left: 2		
Left: 3		
Left: average		
Right: 1		
Right: 2		
Right: 3		
Right: average		

11. Write a procedure for your experiment. Make a list of all the safety precautions you will take. Have your teacher approve your procedure and safety precautions before you begin the experiment.

12. Set up your group's experiment and collect data.

Analyze and Conclude

1. **Summarizing Results** What was your fastest reaction time?

____

2. **Analyzing Data** How does your reaction time when using your dominant hand compare with your reaction time when using your other hand?

____

____

3. **Drawing Conclusions** Why may each hand have a different reaction time? Why may each person have a different reaction time? Compared to earlier trials, was the reaction time in step 8 faster, slower or the same? If the time was faster or slower, hypothesize a reason for the difference.

____

____

____

4. **Predicting Patterns** Compile the data gathered by each pair in your class. Can you identify any trends in the data? (For example, do males and females have the same average reaction times?)

____

____

____

5. **Further Inquiry** Write a new question about reaction times that could be explored in another investigation.

____

____

____

Name ______________________________ Class ______________ Date ____________

Exploration Lab

OBSERVATION

Investigating Vision

What we know about the world is determined by our senses. For example, most people learn about their surroundings through their sense of sight. The lens of our eye focuses light on the retina, the light-sensitive inner layer. The retina contains two types of photoreceptors—rods and cones. *Rods* respond best to dim light and are sensitive to detecting motion. *Cones* respond best to bright light and are responsible for our ability to see colors. Both rods and cones contain light-sensitive chemicals. Nerves connect photoreceptors to the brain, where impulses are organized. We sense the amplitude of light waves as brightness, the patterning as shapes, and the frequency as color.

In this lab, you will use your sense of vision to give you a better understanding of the potentials for and the limits of sight.

OBJECTIVES

Discover which of your eyes is dominant.

Observe how your pupils respond to light.

Discover areas in your field of vision where light is not perceived (blind spot).

Relate binocular vision to depth perception.

MATERIALS

- coin
- dark handkerchief or cloth
- pencil

Procedure

PART I: DOMINANCE

In the same way that you are right-handed or left-handed, one of your eyes is dominant, or used more than the other. Use the following procedure to find out which of your eyes is dominant.

1. Find a small object at least 10 feet away. Close your left eye, and point to the object. Hold your hand steady, and open both eyes. Notice whether you are still pointing to the object.
2. Repeat step 1, using your right eye. Record the eye for which your finger remained in line with the object. This is your dominant eye.

____

____

PART II: BLIND SPOT

The place where the optic nerve exits the retina has no rods or cones. Light that is focused here cannot be perceived. You do not notice a hole in your field of vision because your brain "fills it in" with information from your other eye and gained during reflexive eye movements. Use the procedure below to find the blind spot in both of your eyes.

3. Hold this page at arm's length. Tilt the page at about a 45° angle, with the top of the page higher than the bottom.

4. Close your right eye, and focus on the solid dot in **Figure 1** with your left eye. Concentrate on looking only at the dot, and slowly move the page closer.

5. Repeat steps 3 and 4, this time closing your left eye and focusing on the cross with your right eye. Describe your observations for both eyes. What happens to the shape (dot or cross) that you are not focusing on as you move the page closer?

____

____

____

____

FIGURE 1

PART III: PUPILLARY REFLEX

Pupillary reflex is the constricting or relaxing of the iris muscles that changes the size of the pupils. The irises are opened and closed by opposing muscles. The change in size of the iris regulates the amount of light that enters the eye. Test your pupillary reflex with a partner.

6. Look into your partner's eyes. Note the color, shape, and size of the pupils and irises. Record your observations on the left side of **Table 1.**

7. Have your partner close both eyes, and place a dark cloth over them. Wait 3 minutes, then remove the cloth while your partner's eyes remain closed. Ask your partner to open his or her eyes. Immediately observe your partner's pupils and irises. Record your observations in **Table 1.**

8. Have your partner do steps 6 and 7 and observe your response.

TABLE 1 OBSERVATIONS OF PUPILS AND IRISES

Original observation	After blindfold is removed

PART IV: BINOCULAR VISION

Having two front-facing eyes (binocular vision) helps us to judge distance. Work with your partner to test this aspect of vision.

9. Hold a coin at your partner's eye level. Move the coin closer to the tip of your partner's nose, then farther away.

10. Observe the way your partner's eyeballs converge and how the size of the pupils adjusts to the distance of the coin.

- What happened to your partner's pupils when you moved the coin closer to your partner's nose?

  ____

  ____

- What happened to your partner's pupils when you moved the coin farther away from your partner's nose?

  ____

  ____

11. Hold a pencil point up at eye level in your right hand (with your elbow slightly bent). Focus on the tip of the pencil.

12. Close your left eye. Attempt to touch the point with the index finger of the left hand, starting with your arm at your side.

13. Repeat steps 11 and 12 with your right eye closed. (Keep the pencil in your right hand.)

14. Switch the pencil to your left hand, and repeat the procedure with each eye, touching the pencil with the right hand. As a control, try the experiment with both eyes open.

- Is your dominant eye more accurate in touching the pencil than your other eye?

  ____

15. Clean up your work area before leaving the lab. Place used paper in a trash can or paper recycling bin.

Analysis

1. **Explaining Events** How does the pupillary reflex help the eye adjust to various brightnesses of light?

____

____

2. **Identifying Relationships** Consider what happened to your partner's eyes when you moved a coin closer to and then farther from the tip of your partner's nose. What is the relationship between pupil size and the distance of the object from the eyes?

____

____

Conclusions

1. **Drawing Conclusions** Explain what advantage binocular vision gave you in touching the pencil.

____

____

2. **Applying Conclusions** Suppose a doctor put drops in your eyes and the drops caused the pupils to open widely. Why might the doctor suggest you wear sunglasses outside?

____

____

Extensions

1. **Research and Communications** As people age, some of their senses become less sensitive. For the sense of vision, does the loss in sensitivity affect amplitude (brightness), frequency (color), patterning (shape), or a combination of these factors? Consult an optometrist to learn the answer.
2. **Research and Communications** Vibrations emitted from a moving object have a different frequency coming toward you than going away. This phenomenon is called the Doppler effect. The Doppler effect is used to measure the speed of objects moving toward or away from us. For example, if a comet is moving toward us, the light is pushed to a shorter wavelength. This is called the blue shift. Use library resources to discover what color shift we would see from Earth as a comet moved away from us.

Inquiry Lab

CONSUMER

Collecting Data Through a Survey

A survey is a well-planned, random inquiry, usually performed using a questionnaire. The questionnaire can be a form for people to fill out, or it can be a personal or telephone interview. Most surveys are designed to obtain specific responses from as many people as possible. Those responses may help a surveyer determine a population's opinions or attitudes about a subject of interest. In this lab, you will conduct a survey and analyze information about a population's attitudes and behavior regarding recycling.

OBJECTIVES

Survey members of a community about recycling.

Identify surveying techniques.

Interpret survey results.

Evaluate the usefulness of a survey.

MATERIALS

- pencil
- paper

Procedure

1. Form cooperative teams of four.
2. Your teacher may direct you to use the survey form on the next page. If you design your own questionnaire, use a plain sheet of paper to list your questions, and create your data sheet using the form on the next page as an example. Have your teacher approve your list of questions before you proceed. Obtain a copy of the survey form you have created for each member of your team.
3. Choose two age groups of people to interview. For example, one group can be age 18 and younger, and the other group can be age 19 and older.
4. Have each member of your team interview five people from each group within the community. Tally their answers by placing a check (√) in the appropriate box on the survey form.
5. Combine your team's data. For the younger group, find the total number of responses for each answer box. Do the same for the older group.
6. Find the percentages for each answer box by dividing the total number in the box by the total number of people interviewed in each age group.
7. Combine the data for both age groups, and find the percentages for each answer.

RECYCLING SURVEY

Question	Yes	No	Undecided	N/A
1. Do you know what the term "recycling" means?				
2. Do you think recycling is important?				
3. Do you recycle plastic?				
4. Do you recycle aluminum?				
5. Do you recycle glass?				
6. Do you recycle paper?				
7. Do you earn money from recycling?				
8. If you do not recycle, do you think you should?				
9. Do you think the city should provide curbside recycling service?				
10. Is going to the recycling center inconvenient?				

Analysis

1. **Examining Data** Does one group tend to recycle more than the other group? Explain your answer.

____

____

2. **Analyzing data** What percentage of all the persons interviewed participate in recycling activities?

____

____

____

3. **Analyzing Data** How many recyclers recycle more than one item?

____

____

4. **Identifying Relationships** Does one group think recycling is important more than the other group? How does this correlate to your answer to question 1?

Conclusions

1. **Drawing Conclusions** Explain how the data from a survey might be biased, and how bias might be avoided.

2. **Evaluating Methods** Which of the questions on the survey form in this lab might be a "leading" question that invites a certain response? How can it be restated to be a better question?

3. **Making Predictions** Which will give the most information—a yes/no questionnaire or one that asks for opinions? Which will provide data that is easier to analyze?

Extension

Research and Communications Discover another method for identifying the characteristics of a population. Use the Internet to find out how ecologists sample populations of organisms to determine numbers, health, and diversity in a community.

Name ______________________________ Class ______________ Date ____________

Data Lab

DATASHEET FOR IN-TEXT LAB

Analyzing Changes During a Nerve Impulse

Background

The graph below illustrates changes that occur in the membrane potential of a neuron during an action potential. Use the graph to answer the following questions.

Action Potential

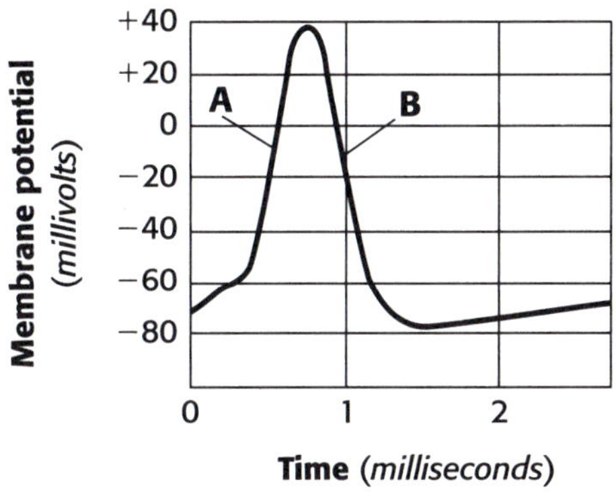

Analysis

1. **Determine** about how long an action potential lasts.

   The action potential lasts about 1.5 milliseconds.

2. **State** whether voltage-gated sodium channels are open or closed at point *A*.

   Voltage-gated sodium channels are open at point *A*.

3. **State** whether voltage-gated potassium channels are open or closed at point *B*.

   Voltage-gated potassium channels are open at point *B*.

4. **Critical Thinking**
   Recognizing Relationships What causes the membrane potential to become less negative at point *A*?

   an influx of positively charged sodium ions

5. **Critical Thinking**
   Recognizing Relationships What causes the membrane potential to become more negative at point *B*?

   an outflow of positively charged potassium ions

Name ______________________________ Class ______________ Date ____________

Quick Lab

DATASHEET FOR IN-TEXT LAB

Demonstrating Your Blind Spot

The blind spot in your visual field corresponds to the site where the optic nerve exits the back of the eye. There are no photoreceptors at the site where the optic nerve exits. Use the procedure below to demonstrate your blind spot.

MATERIALS

- unlined 3 × 5 index card
- pencil

Procedure

1. On the index card, draw an *X* about 1 in. from the left side of the card. Draw an *O* about the same size 3 in. to the right of the *X*.
2. Hold your index card in front of you at arm's length. Close your right eye and stare at the *O* with your left eye. Slowly move the card toward you while continuing to stare at the *O* until the *X* disappears from view.

Analysis

1. **Name** the two kinds of photoreceptors found in the retina.

   rods and cones

2. **Propose** why you cannot see images that fall on the site where the optic nerve exits the eye.

   There are no photoreceptors (rods and cones) at that spot.

3. **Critical Thinking**
   Relating Concepts What is the relationship between the structure of the retina and the disappearance of the *X* on the index card?

   The *X* disappears when it is in the part of the visual field mediated by the part of the retina where there are no photoreceptors.

Name ______________________ Class ______________ Date ______________

Exploration Lab

DATASHEET FOR IN-TEXT LAB

Calculating Reaction Times

SKILLS

- Measuring
- Calculating

OBJECTIVES

- **Determine** human reaction times.
- **Design** an experiment that measures changes in reaction times.

MATERIALS

- meterstick

Before You Begin

When you want to move your hand, your brain must send a message all the way to the muscles in your arms. How long does that take? In this exercise, you will work with a partner to see how quickly you can react. In this lab, you will investigate reaction times and design an experiment to investigate influences on reaction times.

1. You will be using the data table provided to record your data.
2. Write a hypothesis about an influence on reaction times. (For example: "People who have eaten breakfast have faster reaction times than people who have not eaten breakfast.")

   Hypotheses may vary, but should be reasonable and testable. Sample answer: the dominant hand will have a faster reaction time than the non-dominant hand.

Procedure

PART A: CALCULATING REACTION TIMES

1. Sit in a chair and have a partner stand facing you while holding a meterstick in a vertical position.
2. Hold your thumb about 3 cm from your fingers near the bottom end of the stick. The meterstick should be positioned to fall between your thumb and fingers.
3. Tell your partner to let go of the meterstick without warning.
4. When your partner releases the meterstick, catch the stick by pressing your thumb and fingers together. Your partner should be ready to catch the top of the meterstick if it begins to tip over.

Name ______________________ Class ______________ Date ______________

Calculating Reaction Times *continued*

5. Record the number of centimeters the stick dropped before you caught it. The distance that the meterstick falls before you catch it can be used to evaluate your reaction time.
6. Repeat the procedure several times, and calculate the average number of centimeters.
7. Try this procedure with your other hand.
8. Close your eyes and have your partner say "now," when the stick is released.
9. Exchange places with your partner, and repeat the procedure.

PART B: DESIGNING YOUR OWN EXPERIMENT

10. Work with the members of your lab group to explore one of the hypotheses written in the **Before You Begin** section of this lab.

You Choose

As you design your experiment, decide the following:

- **a.** what hypothesis you will explore
- **c.** how you will test the hypothesis
- **d.** what the controls will be
- **e.** how many trials to perform
- **f.** what data to record in your data table

Data Table

Hand: trial number	Subject 1 reaction time (s)	Subject 2 reaction time (s)
Left: 1		
Left: 2		
Left: 3		
Left: average		
Right: 1		
Right: 2		
Right: 3		
Right: average		

11. Write a procedure for your experiment. Make a list of all the safety precautions you will take. Have your teacher approve your procedure and safety precautions before you begin the experiment.
12. Set up your group's experiment and collect data.

Name ______________________ Class ______________ Date ______________

Analyze and Conclude

1. **Summarizing Results** What was your fastest reaction time?

   Answers may vary.

2. **Analyzing Data** How does your reaction time when using your dominant hand compare with your reaction time when using your other hand?

   Answers may vary, but reaction times are often faster in dominant hands.

3. **Drawing Conclusions** Why may each hand have a different reaction time? Why may each person have a different reaction time? Compared to earlier trials, was the reaction time in step 8 faster, slower or the same? If the time was faster or slower, hypothesize a reason for the difference.

   Answers may vary. Reaction times calculated when the subject has closed eyes will probably be slower. Communication is an additional step, and sending and receiving the information that the stick is dropping takes time.

4. **Predicting Patterns** Compile the data gathered by each pair in your class. Can you identify any trends in the data? (For example, do males and females have the same average reaction times?)

   Answers may vary.

5. **Further Inquiry** Write a new question about reaction times that could be explored in another investigation.

   Answers may vary. Sample answer: Would training be a method of improving reaction time? Do students who eat breakfast have faster reaction times than students who skip breakfast? Can caffeine increase reaction time?

Exploration Lab

OBSERVATION

Investigating Vision

Teacher Notes

TIME REQUIRED One 45-minute period

SKILLS ACQUIRED

Collecting data
Experimenting
Interpreting
Inferring
Identifying patterns
Predicting

RATING Easy ← 1 2 3 4 → Hard

Teacher Prep–1
Student Setup–1
Concept Level–2
Cleanup–1

THE SCIENTIFIC METHOD

Make Observations Students make observations in each of the four parts of the Procedure.

Ask Questions Analysis questions 1 and 2 prompt students to ask questions about their observations that will lead them to a greater understanding of vision.

Draw Conclusions Conclusions question 1 asks students to draw conclusions about the advantages of binocular vision.

DISPOSAL

Have students place used paper in a trash can or paper recycling bin.

TIPS AND TRICKS

This lab works best in groups of two students.

It may be helpful if you try each of the procedures beforehand so that you can give hints to students needing assistance.

If students have difficulty finding their dominant eye, have them look through the viewfinder of a camera or microscope. The eye used to sight the image is dominant.

Name ______________________ Class ______________ Date ____________

Exploration Lab

OBSERVATION

Investigating Vision

What we know about the world is determined by our senses. For example, most people learn about their surroundings through their sense of sight. The lens of our eye focuses light on the retina, the light-sensitive inner layer. The retina contains two types of photoreceptors—rods and cones. *Rods* respond best to dim light and are sensitive to detecting motion. *Cones* respond best to bright light and are responsible for our ability to see colors. Both rods and cones contain light-sensitive chemicals. Nerves connect photoreceptors to the brain, where impulses are organized. We sense the amplitude of light waves as brightness, the patterning as shapes, and the frequency as color.

In this lab, you will use your sense of vision to give you a better understanding of the potentials for and the limits of sight.

OBJECTIVES

Discover which of your eyes is dominant.

Observe how your pupils respond to light.

Discover areas in your field of vision where light is not perceived (blind spot).

Relate binocular vision to depth perception.

MATERIALS

- coin
- dark handkerchief or cloth
- pencil

Procedure

PART I: DOMINANCE

In the same way that you are right-handed or left-handed, one of your eyes is dominant, or used more than the other. Use the following procedure to find out which of your eyes is dominant.

1. Find a small object at least 10 feet away. Close your left eye, and point to the object. Hold your hand steady, and open both eyes. Notice whether you are still pointing to the object.
2. Repeat step 1, using your right eye. Record the eye for which your finger remained in line with the object. This is your dominant eye.

   Answers will vary.

Name ______________________ Class ______________ Date ______________

PART II: BLIND SPOT

The place where the optic nerve exits the retina has no rods or cones. Light that is focused here cannot be perceived. You do not notice a hole in your field of vision because your brain "fills it in" with information from your other eye and gained during reflexive eye movements. Use the procedure below to find the blind spot in both of your eyes.

3. Hold this page at arm's length. Tilt the page at about a 45° angle, with the top of the page higher than the bottom.

4. Close your right eye, and focus on the solid dot in **Figure 1** with your left eye. Concentrate on looking only at the dot, and slowly move the page closer.

5. Repeat steps 3 and 4, this time closing your left eye and focusing on the cross with your right eye. Describe your observations for both eyes. What happens to the shape (dot or cross) that you are not focusing on as you move the page closer?

Answers will vary, but with the right eye closed, the cross disappears from view at about half an arm's length. It then reappears almost immediately. As the page is brought closer, some students may observe that the dot disappears. With the left eye closed, the dot disappears first.

FIGURE 1

PART III: PUPILLARY REFLEX

Pupillary reflex is the constricting or relaxing of the iris muscles that changes the size of the pupils. The irises are opened and closed by opposing muscles. The change in size of the iris regulates the amount of light that enters the eye. Test your pupillary reflex with a partner.

6. Look into your partner's eyes. Note the color, shape, and size of the pupils and irises. Record your observations on the left side of **Table 1.**

7. Have your partner close both eyes, and place a dark cloth over them. Wait 3 minutes, then remove the cloth while your partner's eyes remain closed. Ask your partner to open his or her eyes. Immediately observe your partner's pupils and irises. Record your observations in **Table 1.**

8. Have your partner do steps 6 and 7 and observe your response.

Name ______________________ Class ______________ Date ____________

Investigating Vision *continued*

TABLE 1 OBSERVATIONS OF PUPILS AND IRISES

Original observation	After blindfold is removed
Answers will vary.	**Answers will vary, but pupils should be larger and then become smaller as they are exposed to more light.**

PART IV: BINOCULAR VISION

Having two front-facing eyes (binocular vision) helps us to judge distance. Work with your partner to test this aspect of vision.

9. Hold a coin at your partner's eye level. Move the coin closer to the tip of your partner's nose, then farther away.

10. Observe the way your partner's eyeballs converge and how the size of the pupils adjusts to the distance of the coin.

- What happened to your partner's pupils when you moved the coin closer to your partner's nose?

  As the coin moved closer to the face, the pupils converge and grow smaller in size.

- What happened to your partner's pupils when you moved the coin farther away from your partner's nose?

  As the coin moved away from the center of the face, the pupils grow larger in size.

11. Hold a pencil point up at eye level in your right hand (with your elbow slightly bent). Focus on the tip of the pencil.

12. Close your left eye. Attempt to touch the point with the index finger of the left hand, starting with your arm at your side.

13. Repeat steps 11 and 12 with your right eye closed. (Keep the pencil in your right hand.)

14. Switch the pencil to your left hand, and repeat the procedure with each eye, touching the pencil with the right hand. As a control, try the experiment with both eyes open.

- Is your dominant eye more accurate in touching the pencil than your other eye?

  The dominant eye should be more accurate, but both eyes may be equal.

15. Clean up your work area before leaving the lab. Place used paper in a trash can or paper recycling bin.

Name ______________________ Class ______________ Date ____________

Investigating Vision *continued*

Analysis

1. **Explaining Events** How does the pupillary reflex help the eye adjust to various brightnesses of light?

   High amplitude (bright) light will cause the iris muscles to constrict the pupil. Low-amplitude (dim) light causes the muscles to open the pupil.

2. **Identifying Relationships** Consider what happened to your partner's eyes when you moved a coin closer to and then farther from the tip of your partner's nose. What is the relationship between pupil size and the distance of the object from the eyes?

   The nearer the object is, the smaller the pupils. The farther the object is, the larger the pupils.

Conclusions

1. **Drawing Conclusions** Explain what advantage binocular vision gave you in touching the pencil.

   It should be easier to touch the pencil with both eyes open because of binocular vision, which gives better depth perception.

2. **Applying Conclusions** Suppose a doctor put drops in your eyes and the drops caused the pupils to open widely. Why might the doctor suggest you wear sunglasses outside?

   The pupils regulate the amount of light that comes into the eye. The sunglasses would protect the eyes from taking in too much bright sunlight.

Extensions

1. **Research and Communications** As people age, some of their senses become less sensitive. For the sense of vision, does the loss in sensitivity affect amplitude (brightness), frequency (color), patterning (shape), or a combination of these factors? Consult an optometrist to learn the answer.

2. **Research and Communications** Vibrations emitted from a moving object have a different frequency coming toward you than going away. This phenomenon is called the Doppler effect. The Doppler effect is used to measure the speed of objects moving toward or away from us. For example, if a comet is moving toward us, the light is pushed to a shorter wavelength. This is called the blue shift. Use library resources to discover what color shift we would see from Earth as a comet moved away from us.

Inquiry Lab

CONSUMER

Collecting Data Through a Survey

Teacher Notes

TIME REQUIRED Two 45-minute periods; students will collect data over a period of several days outside of class.

SKILLS ACQUIRED

Collecting data
Identifying patterns
Inferring
Interpreting
Organizing and analyzing data

RATING Easy ← 1 2 3 4 → Hard

Teacher Prep–1
Student Setup–2
Concept Level–3
Cleanup–1

THE SCIENTIFIC METHOD

Analyze the Results Analysis questions 2 and 3 ask students to analyze their results.

Draw Conclusions Conclusions question 1 asks students to draw conclusions from their data.

MATERIALS

Materials for this lab activity can be purchased from WARD'S. See the *Master Materials List* for ordering instructions.

TIPS AND TRICKS

Preparation

This lab works best in groups of four students.

You may wish to have a discussion of survey techniques, and have students develop their own survey questions as a homework assignment. If so, have students submit their questions a day or two ahead of time so you can duplicate enough survey forms to supply each student before the first lab period.

Procedure

Students may use the survey form in the lab and make tally marks in the response blocks.

A survey works best when many subjects are interviewed. Encourage students to survey people outside their own family members if possible. If students decide to interview people "on the street," caution them not to do so alone. They should work only as a team in a busy public area.

Name ______________________ Class ______________ Date ______________

Inquiry Lab

CONSUMER

Collecting Data Through a Survey

A survey is a well-planned, random inquiry, usually performed using a questionnaire. The questionnaire can be a form for people to fill out, or it can be a personal or telephone interview. Most surveys are designed to obtain specific responses from as many people as possible. Those responses may help a surveyer determine a population's opinions or attitudes about a subject of interest. In this lab, you will conduct a survey and analyze information about a population's attitudes and behavior regarding recycling.

OBJECTIVES

Survey members of a community about recycling.

Identify surveying techniques.

Interpret survey results.

Evaluate the usefulness of a survey.

MATERIALS

- pencil
- paper

Procedure

1. Form cooperative teams of four.
2. Your teacher may direct you to use the survey form on the next page. If you design your own questionnaire, use a plain sheet of paper to list your questions, and create your data sheet using the form on the next page as an example. Have your teacher approve your list of questions before you proceed. Obtain a copy of the survey form you have created for each member of your team.
3. Choose two age groups of people to interview. For example, one group can be age 18 and younger, and the other group can be age 19 and older.
4. Have each member of your team interview five people from each group within the community. Tally their answers by placing a check (√) in the appropriate box on the survey form.
5. Combine your team's data. For the younger group, find the total number of responses for each answer box. Do the same for the older group.
6. Find the percentages for each answer box by dividing the total number in the box by the total number of people interviewed in each age group.
7. Combine the data for both age groups, and find the percentages for each answer.

Name ______________________ Class ______________ Date ____________

Collecting Data Through a Survey *continued*

RECYCLING SURVEY

Question	Yes	No	Undecided	N/A
1. Do you know what the term "recycling" means?				
2. Do you think recycling is important?				
3. Do you recycle plastic?				
4. Do you recycle aluminum?				
5. Do you recycle glass?				
6. Do you recycle paper?				
7. Do you earn money from recycling?				
8. If you do not recycle, do you think you should?				
9. Do you think the city should provide curbside recycling service?				
10. Is going to the recycling center inconvenient?				

Analysis

1. **Examining Data** Does one group tend to recycle more than the other group? Explain your answer.

   Answers will depend on the data, but students should be able to tell at a glance at their percentages if one group recycles more.

2. **Analyzing data** What percentage of all the persons interviewed participate in recycling activities?

   Students should add the percentages that indicate the respondent recycles anything, regardless of other answers, and divide by the total number interviewed.

3. **Analyzing Data** How many recyclers recycle more than one item?

   Students should add responses that indicate that a person recycles at least two items and find the percentage of the total number interviewed.

Name ______________________ Class ______________ Date ____________

4. Identifying Relationships Does one group think recycling is important more than the other group? How does this correlate to your answer to question 1?

Students will probably find that the group that recycles more also thinks recycling is more important in higher percentages. If the reverse is true according to their data, students should provide a possible explanation.

Conclusions

1. Drawing Conclusions Explain how the data from a survey might be biased, and how bias might be avoided.

The survey could be biased by asking biased questions or by a poorly randomized group of interviewees. Bias can be avoided by making certain that the survey questions do not invite a certain response, that the group of interviewees is randomly selected, and that the group is as large as possible.

2. Evaluating Methods Which of the questions on the survey form in this lab might be a "leading" question that invites a certain response? How can it be restated to be a better question?

Question 10 is a leading question. Accept restated questions that avoid a negative word, such as *inconvenient*. For example, "Do you visit a recycling center often?"

3. Making Predictions Which will give the most information—a yes/no questionnaire or one that asks for opinions? Which will provide data that is easier to analyze?

An open-ended questionnaire provides more in-depth information, but the yes/no questionnaire is much easier to analyze. An open-ended questionnaire could have as many different answers as there are interviewees.

Extension

Research and Communications Discover another method for identifying the characteristics of a population. Use the Internet to find out how ecologists sample populations of organisms to determine numbers, health, and diversity in a community.

Answer Key

Directed Reading

SECTION: NEURONS AND NERVE IMPULSES

1. d
2. a
3. b
4. c
5. e
6. The membrane potential is caused by the relative concentration of ions inside and outside the cell, the ability of the ions to diffuse across the cell membrane, and the electrical charge of the ions.
7. The resting potential is negative because the concentration of sodium ions is greater outside the cell than it is inside the cell, and the concentration of potassium ions is greater inside the cell than it is outside the cell. Because voltage-gated sodium ions are closed at resting potential and voltage-gated potassium ions are open, only positively-charged potassium ions diffuse out of the cell. Neurons also contain negatively charged proteins that are too large to exit the cell.
8. An action potential, or nerve impulse, is a local reversal of polarity—from a negative charge to a positive charge—inside a neuron.
9. 4
10. 3
11. 1
12. 2
13. 2
14. 3
15. 1
16. 4

SECTION: STRUCTURES OF THE NERVOUS SYSTEM

1. The central nervous system (CNS) consists of the brain and spinal cord. It is the control center of the body. The CNS interprets and responds to information. The peripheral nervous system contains sensory and motor neurons that carry information to and from the CNS.
2. Sensory neurons carry information from the sense organs to the CNS. Motor neurons send commands from the CNS to muscles and other organs.
3. g
4. c
5. f
6. b
7. d
8. a
9. e
10. somatic
11. parasympathetic
12. sympathetic

SECTION: SENSORY SYSTEMS

1. A sensory receptor is a specialized neuron that detects sensory stimuli and then converts the stimuli into electrical signals that can be interpreted by the brain.
2. thermoreceptors, mechanoreceptors, and pain receptors
3. Pain is an important sensation because it indicates that something is wrong in the body.
4. mechanoreceptors
5. frontal, occipital, temporal, and parietal
6. retina
7. rods, cones
8. optic nerve
9. The cochlea is a fluid-filled chamber within the inner ear that contains hair cells. The hair cells are stimulated when sound waves enter the cochlea, sending signals to the brain.
10. Semicircular canals are fluid-filled chambers in the inner ear that contain hair cells. When the head moves, the hair cells are stimulated and send electrical signals to the brain, enabling the brain to determine the orientation and position of the head.
11. salts
12. olfactory

SECTION: DRUGS AND THE NERVOUS SYSTEM

1. A psychoactive drug is a drug that alters the functioning of the central nervous system.
2. depressants
3. disorientation, confusion, and memory loss
4. Marijuana and hashish contain THC. Marijuana and hashish can cause impaired judgment, short-term memory loss, and loss of motivation.
5. coma and respiratory failure
6. c
7. b
8. d
9. a
10. Cocaine blocks the reuptake of dopamine molecules by interfering with reuptake receptors on presynaptic neurons. As a result, excess dopamine remains in the synaptic cleft, overstimulating postsynaptic cells.
11. Cocaine use leads to a reduction in the number of dopamine receptors on the postsynaptic cells that it affects. Addiction occurs because more cocaine must be taken to maintain adequate stimulation of the postsynaptic cells.
12. A depressant is a drug that decreases the activity of the central nervous system.
13. Alcoholism can lead to cirrhosis of the liver, malnutrition, abnormalities in the circulatory system, and inflammation of the stomach lining.
14. nicotine
15. acetylcholine
16. tars or mutagens
17. enkephalins
18. opium

Active Reading

SECTION: NEURONS AND NERVE IMPULSES

1. by transmitting nerve impulses to body tissues and organs, including muscles, glands, and other neurons
2. Dendrites receive information from other cells.
3. collects information from dendrites, relays this information to other parts of the neuron, and maintains the general functioning of the neuron
4. a long extension of the cytoplasm that conducts nerve impulses
5. Nerves are bundles of axons.
6. c

SECTION: STRUCTURES OF THE NERVOUS SYSTEM

1. b	11. a
2. c	12. b
3. a	13. a
4. a	14. c
5. c	15. b
6. a	16. a
7. b	17. b
8. a	18. c
9. c	19. a
10. c	20. d

SECTION: SENSORY SYSTEMS

1. The ear enables you to hear and helps you maintain your balance.
2. Energy in sound waves is changed to electrical signals in the ear.
3. The tympanic membrane vibrates when sound waves pass through the ear canal and strike the eardrum.
4. a fluid-filled chamber within the inner ear
5. The bones of the middle ear, which are the hammer, anvil, and stirrup, transfer vibrations from the eardrum to the cochlea.
6. The hair cells generate nerve impulses in the auditory nerve.
7. fluid-filled chambers in the inner ear that contain hair cells
8. Hair cells in the semicircular canals respond to changes in head position with respect to gravity. The hair cells generate electrical signals and send these signals to the brain. The brain interprets these signals to determine the orientation and position of the head.
9. d

SECTION: DRUGS AND THE NERVOUS SYSTEM

1. A depressant is a drug that generally decreases the activity of the central nervous system.

2. Reaction times increase and coordination, judgment, and speech become impaired.
3. Alcohol is absorbed into the blood through the stomach and small intestine.
4. a

Vocabulary Review

1. d
2. f
3. b
4. e
5. a
6. c
7. i
8. g
9. h
10. motor neuron
11. cerebrum
12. central nervous system
13. thalamus
14. spinal cord
15. peripheral nervous system
16. sensory neuron
17. brain
18. hypothalamus
19. interneurons
20. brain stem
21. cerebellum
22. reflex
23. f
24. a
25. c
26. g
27. d
28. b
29. e
30. tolerance
31. depressant
32. stimulant
33. psychoactive drugs
34. withdrawal
35. addiction

Science Skills

INTERPRETING DATA

1. The larger the diameter of the axon, the faster it conducts impulses.
2. Student answers will vary but may include the idea that sensory impulses from the skin, such as touch, pressure, and temperature, are carried by A fibers. Motor impulses to skeletal muscles are also carried by A fibers. These types of impulses relay information about the external surroundings and how the body will respond to external stimuli. The speed with which these impulses are carried could reduce the incidence of injury to the body by allowing for a quick reaction to a stimulus.
3. Student answers will vary. Sample answer: Two factors that influence the speed with which an impulse is conducted are the diameter of the axon and whether or not it is myelinated. In the absence of myelination, increasing the diameter of the axon will increase the speed of impulse conduction. The large axons in the organisms are responsible for the escape or withdrawl reflexes.

INTERPRETING SCIENTIFIC ILLUSTRATIONS

4. axon terminal
5. neurotransmitter molecule
6. drug molecule
7. reuptake receptor
8. receptor protein
9. The drug molecule binds to the reuptake receptor that would normally remove the neurotransmitter molecules from the synapse and end the impulse. As a result, the impulse continues and the postsynaptic neuron is overstimulated.

Concept Mapping

1. neurons
2. central nervous system
3. peripheral nervous system
4. synapses
5. somatic nervous system
6. spinal cord
7. interneurons
8. motor neurons
9. sympathetic division or parasympathetic division
10. sympathetic division or parasympathetic division

Critical Thinking

1. e
2. d
3. a
4. c
5. f
6. b
7. c
8. b
9. f
10. e
11. a
12. d
13. e
14. d
15. a
16. b
17. c
18. d, j
19. i, a
20. k, c
21. l, e
22. b, h
23. g, f
24. c
25. d
26. c
27. a

Test Prep Pretest

1. b
2. a
3. d
4. a
5. c
6. d
7. a
8. d
9. c
10. b
11. c
12. resting potential
13. action potential
14. neurotransmitter, cleft
15. excite, inhibit
16. potassium, negative
17. gray, white
18. peripheral
19. cochlea
20. brain
21. temporal lobe
22. psychoactive drugs
23. depressant
24. The somatic nervous system contains motor neurons that function in conscious activities, such as contraction of voluntary muscles. The autonomic nervous system contains motor neurons that function in unconscious activities, such as contraction of involuntary muscles and stimulation of glandular secretions.
25. sugars (sweet), acids (sour), alkaloids (bitter), and salts (salty)

Quiz

SECTION: NEURONS AND NERVE IMPULSES

1. c
2. e
3. d
4. g
5. a
6. f
7. b
8. c
9. b
10. d

SECTION: STRUCTURES OF THE NERVOUS SYSTEM

1. c
2. b
3. d
4. c
5. e
6. g
7. b
8. f
9. d
10. a

SECTION: SENSORY SYSTEMS

1. b
2. c
3. c
4. b
5. d
6. c
7. b
8. d
9. e
10. a

SECTION: DRUGS AND THE NERVOUS SYSTEM

1. d
2. a
3. a
4. d
5. e
6. f
7. c
8. b
9. d
10. a

Chapter Test (General)

1. a
2. a
3. c
4. d
5. c
6. b
7. d
8. b
9. c
10. d
11. b
12. a
13. c
14. d
15. b
16. a
17. c
18. a
19. d
20. b

Chapter Test (Advanced)

1. b
2. a
3. c
4. d
5. c
6. a
7. a

8. b
9. d
10. b
11. d
12. a
13. c
14. b
15. taste bud
16. cell bodies, axons
17. sodium, sodium
18. mechanoreceptors
19. olfactory
20. temporal
21. Signals are transmitted at a synapse by neurotransmitters. When a nerve impulse reaches the end of an axon, neurotransmitters are released from the presynaptic neuron into the synaptic cleft. The neurotransmitters move across the synapse and bind to receptors on the postsynaptic cell. This either excites or inhibits the postsynaptic cell.
22. Most sensory and motor processing occurs in the outer layer of the cerebrum, called the cerebral cortex. The capacity for learning, memory, perception, and intellectual function resides in the cerebrum.
23. Addiction is a physiological response caused by the use of a drug that alters the normal functioning of neurons and synapses. Once a neuron or synapse has been altered by a drug, it cannot function normally unless the drug is present.
24. Sound waves enter the ear canal and cause the eardrum to vibrate. Three small bones in the middle ear—the hammer, anvil, and stirrup—transfer the vibrations to a membrane in the fluid-filled cochlea in the inner ear. The vibrations reach the membrane and stimulate hair cells, which then send nerve impulses to the brain stem via the auditory nerve. There the thalamus relays the information to the temporal lobe of the cerebral cortex, where the auditory information is processed.
25. Cocaine blocks the reuptake of dopamine by the presynaptic neuron. The excess dopamine overstimulates the postsynaptic cell, which thereby decreases the number of dopamine receptors. This causes the postsynaptic cell to become less sensitive, requiring more cocaine for adequate stimulation.

Lesson Plan

Section: Neurons and Nerve Impulses

Pacing

Regular Schedule:	**with lab(s):** N/A	**without lab(s):** 3 days
Block Schedule:	**with lab(s):** N/A	**without lab(s):** 1 1/2 days

Objectives

1. Analyze the structure and function of neurons.
2. Describe how the resting potential is established in a neuron.
3. Sequence the steps of a nerve impulse.
4. List the events that occur in synaptic transmission of a nerve impulse.

National Science Education Standards Covered

UNIFYING CONCEPTS AND PROCESSES

UCP1: Systems, order, and organization

UCP5: Form and function

SCIENCE AS INQUIRY

SI1: Abilities necessary to do scientific inquiry

SI2: Understandings about scientific inquiry

LIFE SCIENCE: THE CELL

LSCell1: Cells have particular structures that underlie their functions.

LSCell2: Most cell functions involve chemical reactions.

LSCell4: Cell functions are regulated.

LIFE SCIENCE: MATTER, ENERGY, AND ORGANIZATION IN LIVING SYSTEMS

LSMat4: The complexity and organization of organisms accommodates the need for obtaining, transforming, transporting, releasing, and eliminating the matter and energy used to sustain the organism.

LIFE SCIENCE: BEHAVIOR OF ORGANISMS

LSBeh1: Multicellular animals have nervous systems that generate behavior.

LSBeh2: Organisms have behavioral responses to internal changes and to external stimuli.

KEY
SE = Student Edition **TE** = Teacher Edition
CRF = Chapter Resource File

Block 1

CHAPTER OPENER *(45 minutes)*

_ **Quick Review,** SE. Students answer questions covered in previous sections of the textbook as preparation for the chapter content. **(GENERAL)**

_ **Reading Activity,** SE. Before students begin this section, ask them whether they agree or disagree with statements written on the board. Have students save their lists for later use. **(GENERAL)**

_ **Using the Figure,** TE. Students answer questions about the chapter opener photograph. **(GENERAL)**

_ **Opening Activity**, Nervous System Function, TE. Ask students to brainstorm the functions of the nervous system. Write their responses on the board, and group them in the following categories: (1) monitoring internal and external environments; (2) coordinating bodily activities; and (3) intellectual function and memory. **(GENERAL)**

Block 2

FOCUS *(5 minutes)*

_ **Bellringer Transparency.** Use this transparency as students enter the classroom and find their seats. **(GENERAL)**

MOTIVATE *(10 minutes)*

_ **Activity**, Information Transfer, TE. Discuss the similarities and differences between the game "Telephone" and nerve impulses as they pass through the body. **(BASIC)**

TEACH *(30 minutes)*

_ **Teaching Transparency, Section Outline.** Use this transparency to give students a framework for the information in this section. **(GENERAL)**

_ **Teaching Transparency, Structure of a Neuron.** Use this transparency to describe the basic structure and function of a neuron. Trace the path an impulse travels. Point out that the axon is covered with an intermittent myelin sheath and that the impulse skips from node to node. When the impulse reaches the axon's tip, neurotransmitters released into the synapse carry the signal to the adjacent cell. **(GENERAL)**

_ **Data Lab,** Analyzing Changes During a Nerve Impulse, SE. Students study a graph illustrating changes in the membrane potential of a neuron during an action potential and answer questions. **(GENERAL)**

_ **Datasheets for In-Text Labs, Analyzing Changes During Nerve Impulse, CRF.**

_ **Teaching Transparency, Conduction of a Nerve Impulse.** Use this transparency to describe the propagation of a nerve impulse. Emphasize that the concentration of sodium is higher outside the cell than inside the cell. Point out that as the nerve impulse travels along the axon of one neuron, sodium ions reach into the cell, and temporarily reverse the voltage in a small patch of membrane. **(GENERAL)**

HOMEWORK

_ **Active Reading Worksheet, Neurons and Nerve Impulses, CRF.** Students read a passage related to the section topic and answer questions. **(GENERAL)**

_ **Directed Reading Worksheet, Neurons and Nerve Impulses, CRF.** Students complete the exercises in this worksheet to help them understand the material as they read the section. **(BASIC)**

Block 3

TEACH *(25 minutes)*

_ **Teaching Tip**, Summarizing an Action Potential, TE. Students summarize in writing the events involved in a neuron's propagation of an action potential. **(GENERAL)**

_ **Skill Builder**, Math Skills, TE. Students calculate how many neurons could be stimulated if 1,000 neurons forms synapses with another 1,000 neurons.

_ **Teaching Transparency, Synaptic Transmission.** Use this transparency to describe how signals are transmitted between a neuron and another cell. Explain that neurotransmitter molecules are released from the presynaptic neuron and they either excite or inhibit the postsynaptic cell by binding to it. **(GENERAL)**

CLOSE *(20 minutes)*

_ **Reteaching,** TE. Students use clay to build a model of a neuron and a synapse. **(BASIC)**

_ **Alternative Assessment**, TE. Students write questionss for a trivia-type game to review the section's objectives. **(GENERAL)**

HOMEWORK

_ **Quiz,** TE. Students answer questions that review the section material. **(GENERAL)**

_ **Section Review,** SE. Assign questions 1–5 for review, homework, or quiz. **(GENERAL)**

_ **Quiz, CRF.** This quiz consists of ten multiple choice and matching questions that review the section's main concepts. **(BASIC) Also in Spanish.**

Other Resource Options

- **go.hrw.com.** For worksheets, videos, and other teaching aids related to this chapter, visit the HRW Web site and type in the keyword HX4 NRV.
- **CNN Science in the News, Video Segment 34 Stroke Brain Repair.** This video segment is accompanied by a **Critical Thinking Worksheet**.
- **Biology Interactive Tutor CD-ROM,** Unit 1 Cell Transport and Homeostasis. Students watch animations and other visuals as the tutor explains cell transport and homeostais. Students assess their learning with interactive activities.
- **CNN Student News.** Find the latest news, lesson plans, and activities related to important scientific events at **cnnstudentnews.com**.

Lesson Plan

Section: Structures of the Nervous System

Pacing

Regular Schedule:	**with lab(s):** 4 days	**without lab(s):** 2 days
Block Schedule:	**with lab(s):** 2 days	**without lab(s):** 1 day

Objectives

1. Distinguish between the central nervous system and the peripheral nervous system.
2. Identify the major parts of the brain and their functions.
3. Describe the structure of the spinal cord.
4. Sequence the events of a spinal reflex.
5. Compare the somatic nervous system with the autonomic nervous system.

National Science Education Standards Covered

UNIFYING CONCEPTS AND PROCESSES

UCP1: Systems, order, and organization

UCP5: Form and function

SCIENCE AS INQUIRY

SI1: Abilities necessary to do scientific inquiry

SI2: Understandings about scientific inquiry

LIFE SCIENCE: MATTER, ENERGY, AND ORGANIZATION IN LIVING SYSTEMS

LSMat4: The complexity and organization of organisms accommodates the need for obtaining, transforming, transporting, releasing, and eliminating the matter and energy used to sustain the organism.

LIFE SCIENCE: BEHAVIOR OF ORGANISMS

LSBeh1: Multicellular animals have nervous systems that generate behavior.

LSBeh2: Organisms have behavioral responses to internal changes and to external stimuli.

KEY

SE = Student Edition **TE** = Teacher Edition
CRF = Chapter Resource File

Block 4

FOCUS *(5 minutes)*

_ **Bellringer Transparency.** Use this transparency as students enter the classroom and find their seats. **(GENERAL)**

MOTIVATE *(10 minutes)*

_ **Demonstration**, TE. Have a volunteer help demonstrate reflexive blinking. **(BASIC)**

TEACH *(30 minutes)*

_ **Teaching Transparency, Section Outline.** Use this transparency to give students a framework for the information in this section. **(GENERAL)**

_ **Real Life**, SE. Students discuss whether there should be helmet laws for bicyclists. **(GENERAL)**

_ **Teaching Transparency, Structure of the Human Brain.** Use this transparency to describe the structure of the brain. Point out the cerebrum, cerebellum, and brain stem. **(GENERAL)**

_ **Teaching Transparency, Structure of the Human Spinal Cord.** Use this transparency to describe the structure and function of the spinal cord. Walk students through the cross section of the spinal cord. **(GENERAL)**

HOMEWORK

_ **Directed Reading Worksheet, Structures of the Nervous System, CRF.** Students complete the exercises in this worksheet to help them understand the material as they read the section. **(BASIC)**

_ **Active Reading Worksheet, Structures of the Nervous System, CRF.** Students read a passage related to the section topic and answer questions. **(GENERAL)**

Block 5

TEACH *(30 minutes)*

_ **Demonstration**, TE. Students work in pairs to demonstrate reaction time using a ruler. **(GENERAL)**

_ **Teaching Transparency, Knee-Jerk Reaction.** Use this transparency to describe the the simple neural circuit of a reflex arc. Guide students through the events, beginning with the hammer tap below the knee cap, which stimulates the quadriceps muscle. Tell students that doctors regularly test the knee-jerk reflex to ascertain whether certain nerves and a portion of the spinal cord are functioning properly. **(GENERAL)**

_ **Inclusion Strategies**, TE. Students work with a partner to make observations about how the body keeps a sense of balance.

_ **Integrating Physics and Chemistry**, TE. Relate membrane, resting, and action potentials to potential difference. Then ask students what biological mechanism pushes electrical charges opposite the electric force.

CLOSE *(15 minutes)*

_ **Reteaching,** TE. List the major concepts from this chapter on the board. Have the class review this section to identify examples that can be used to illustrate each of these concepts. **(BASIC)**

_ **Quiz,** TE. Students answer questions that review the section material. **(GENERAL)**

HOMEWORK

_ **Section Review,** SE. Assign questions 1–6 for review, homework, or quiz. **(GENERAL)**

_ **Quiz, CRF.** This quiz consists of ten multiple choice and matching questions that review the section's main concepts. **(BASIC) Also in Spanish.**

Optional Blocks

LAB *(90 minutes)*

_ **Inquiry Lab, Collecting Data Through a Survey, CRF.** Students conduct a survey and analyze information about a population's attitudes and behavior regarding recycling. **(GENERAL)**

Other Resource Options

_ **Alternative Assessment,** TE. Students take turns lightly tapping each other's leg slightly below the knee. Ask students to write a brief report summarizing their observations and the cause of the effect produced. **(GENERAL)**

_ **Internet Connect.** Students can research Internet sources about the Spinal Cord with SciLinks Code HX4168.

_ **go.hrw.com.** For worksheets, videos, and other teaching aids related to this chapter, visit the HRW Web site and type in the keyword HX4 NRV.

_ **CNN Science in the News, Video Segment 34 Stroke Brain Repair.** This video segment is accompanied by a **Critical Thinking Worksheet**.

_ **Biology Interactive Tutor CD-ROM,** Unit 1 Cell Transport and Homeostasis. Students watch animations and other visuals as the tutor explains cell transport and homeostais. Students assess their learning with interactive activities.

Lesson Plan

Section: Sensory Systems

Pacing

Regular Schedule: **with lab(s):** 3 days **without lab(s):** 2 days

Block Schedule: **with lab(s):** 1 1/2 days **without lab(s):** 1 day

Objectives

1. List five types of sensory receptors and the stimuli to which they respond.
2. Identify sites of sensory processing in the brain.
3. Analyze the structure of the eye and its role in the visual system.
4. Describe how the ear detects sound and helps maintain balance.
5. Compare the senses of taste and smell.

National Science Education Standards Covered

UNIFYING CONCEPTS AND PROCESSES

UCP1: Systems, order, and organization

UCP5: Form and function

SCIENCE AS INQUIRY

SI1: Abilities necessary to do scientific inquiry

SI2: Understandings about scientific inquiry

LIFE SCIENCE: THE CELL

LSCell1: Cells have particular structures that underlie their functions.

LSCell2: Most cell functions involve chemical reactions.

LSCell4: Cell functions are regulated.

LIFE SCIENCE: MATTER, ENERGY, AND ORGANIZATION IN LIVING SYSTEMS

LSMat4: The complexity and organization of organisms accommodates the need for obtaining, transforming, transporting, releasing, and eliminating the matter and energy used to sustain the organism.

LIFE SCIENCE: BEHAVIOR OF ORGANISMS

LSBeh2: Organisms have behavioral responses to internal changes and to external stimuli.

KEY
SE = Student Edition **TE** = Teacher Edition
CRF = Chapter Resource File

Block 6

FOCUS *(5 minutes)*

_ **Bellringer Transparency.** Use this transparency as students enter the classroom and find their seats. **(GENERAL)**

MOTIVATE *(10 minutes)*

_ **Activity**, Phosphenes, TE. Have students close their eyes and press lightly on their eyelids until they see an effect. Ask them what they see. Discuss phosphenes and their causes. **(GENERAL)**

TEACH *(30 minutes)*

_ **Teaching Transparency, Section Outline.** Use this transparency to give students a framework for the information in this section. **(GENERAL)**

_ **Teaching Transparency, Types of Sensory Receptors.** Use this transparency to discuss different types of sensory receptors and their locations. **(GENERAL)**

_ **Teaching Transparency, Processing Sites and Lobes of the Cerebral Cortex.** Use this transparency to point out the hemispheres and lobes of the brain. Discuss how the sensory neurons from the different sense organs come together at certain regions of the cerebral cortex. **(GENERAL)**

_ **Activity**, Visual Feedback and Touch, TE. Students twist their arms in an usual position and then try to move fingers as they are pointed to by a partner. **(GENERAL)**

_ **Teaching Transparency, Structure of the Eye.** Use this transparency to discuss the structure of the human eye. Review the major structures involved in vision. Tell students that the human eye is a camera-type eye. It uses a single lens to focus light on a surface containing many photoreceptors. In this type of eye, the retina functions in a manner analogous to a piece of film. **(GENERAL)**

HOMEWORK

_ **Directed Reading Worksheet, Sensory Systems, CRF.** Students complete the exercises in this worksheet to help them understand the material as they read the section. **(BASIC)**

_ **Active Reading Worksheet, Sensory Systems, CRF.** Students read a passage related to the section topic and answer questions. **(GENERAL)**

Block 7

TEACH *(35 minutes)*

- **Integrating Physics and Chemistry**, TE. Review the properties of light. Ask students to identify and describe which property of light occurs as light passes through the lens.
- **Quick Lab,** Demonstrating Your Blind Spot, SE. Students locate their own blind spot. **(GENERAL)**
- **Datasheets for In-Text Labs, Demonstrating Your Blind Spot, CRF.**
- **Inclusion Strategies**, TE. Students work with a partner to make observations on sound detection.
- **Teaching Transparency, Structure of the Ear.** Use this transparency to discuss the structures and functions of the inner ear. Review how the structures in the inner ear sense equilibrium and explain how structures in the inner ear sense sound. Point out that the cuplike shape of the outer ear, the pinnae, captures sound vibrations in the air and funnels the vibrations into the auditory canal. **(GENERAL)**

CLOSE *(10 minutes)*

- **Reteaching,** TE. Assign students to cooperative groups of four. Have the groups choose the sensory system they would least like to lose. Have each group defend its choice. **(BASIC)**
- **Quiz,** TE. Students answer questions that review the section material. **(GENERAL)**

HOMEWORK

- **Alternative Assessment,** TE. Students prepare a diagram that traces sound from its origin to the brain. **(GENERAL)**
- **Section Review,** SE. Assign questions 1–5 for review, homework, or quiz. **(GENERAL)**

Optional Block

LAB *(45 minutes)*

- **Exploration Lab, Investigating Vision, CRF.** Students use their sense of vision to get a fuller understanding of the potentials for and the limits of sight. **(GENERAL)**

Other Resource Options

- **Internet Connect.** Students can research Internet sources about Autonomic Nervous System with SciLinks Code HX4017.
- **Internet Connect.** Students can research Internet sources about Sensory Receptors with SciLinks Code HX4162.

_ **Internet Connect.** Students can research Internet sources about Chemical Senses with SciLinks Code HX4041.

_ **go.hrw.com.** For worksheets, videos, and other teaching aids related to this chapter, visit the HRW Web site and type in the keyword HX4 NRV.

_ **CNN Science in the News, Video Segment 34 Stroke Brain Repair.** This video segment is accompanied by a **Critical Thinking Worksheet**.

_ **Biology Interactive Tutor CD-ROM,** Unit 1 Cell Transport and Homeostasis. Students watch animations and other visuals as the tutor explains cell transport and homeostais. Students assess their learning with interactive activities.

_ **CNN Student News.** Find the latest news, lesson plans, and activities related to important scientific events at **cnnstudentnews.com**.

Lesson Plan

Section: Drugs and the Nervous System

Pacing

Regular Schedule:	**with lab(s):** 3 days	**without lab(s):** 2 days
Block Schedule:	**with lab(s):** 1 1/2 days	**without lab(s):** 1 day

Objectives

1. Identify types of psychoactive drugs, and describe their effects.
2. Describe how drug addiction develops.
3. Describe effects of commonly abused drugs on the nervous system.

National Science Education Standards Covered

UNIFYING CONCEPTS AND PROCESSES

UCP5: Form and function

SCIENCE AS INQUIRY

SI1: Abilities necessary to do scientific inquiry

SI2: Understandings about scientific inquiry

LIFE SCIENCE: THE CELL

LSCell1: Cells have particular structures that underlie their functions.

LSCell2: Most cell functions involve chemical reactions.

LSCell4: Cell functions are regulated.

LIFE SCIENCE: MATTER, ENERGY, AND ORGANIZATION IN LIVING SYSTEMS

LSMat4: The complexity and organization of organisms accommodates the need for obtaining, transforming, transporting, releasing, and eliminating the matter and energy used to sustain the organism.

LIFE SCIENCE: BEHAVIOR OF ORGANISMS

LSBeh1: Multicellular animals have nervous systems that generate behavior.

LSBeh2: Organisms have behavioral responses to internal changes and to external stimuli.

KEY
SE = Student Edition **TE** = Teacher Edition
CRF = Chapter Resource File

Block 8

FOCUS *(5 minutes)*

_ **Bellringer Transparency.** Use this transparency as students enter the classroom and find their seats. **(GENERAL)**

MOTIVATE *(10 minutes)*

_ **Demonstration**, TE. Demonstrate the effect of secondhand smoke by burning several cigarettes in an ashtray under a fume hood. Place a piece of glass about 5 cm above the cigarettes. After the cigarettes have burned completely, wipe the glass with a white cloth and show the cloth to the class. **(GENERAL)**

TEACH *(30 minutes)*

_ **Teaching Transparency, Section Outline.** Use this transparency to give students a framework for the information in this section. **(GENERAL)**

_ **Teaching Transparency, Psychoactive Drugs of Abuse.** Use this transparency to discuss the classes of commonly abused drugs. **(GENERAL)**

_ **Teaching Tip**, Effects of Drugs, TE. Students choose a class of frequently abused drugs or a specific drug and make a pamphlet about the effects of the drug, the dangers of using the drug, and avenues for seeking help for drug addiction. **(GENERAL)**

_ **Group Activity**, Drugs and Their Effects, TE. Students work in groups to make a table listing different kinds of drugs, their short-term effects, their long-term effects, and their desired effects. Have students categorize he drugs as medicinal or social, legal or illegal, and or prescription or over-the-counter. **(GENERAL)**

HOMEWORK

_ **Directed Reading Worksheet, Drugs and the Nervous System, CRF.** Students complete the exercises in this worksheet to help them understand the material as they read the section. **(BASIC)**

_ **Active Reading Worksheet, Drugs and the Nervous System, CRF.** Students read a passage related to the section topic and answer questions. **(GENERAL)**

_ **Occupational Application Worksheet, Pharmacist, One-Stop Planner.** Students learn what a pharmacist does by filling out several patient profiles and completing a drug inventory form. **(GENERAL)**

_ **Activity**, BAC, TE. Have students research laws for blood alcohol concentration. **(GENERAL)**

Block 9

TEACH *(30 minutes)*

- **Real Life**, SE. Students determine their daily total caffeine intake for one day. **(GENERAL)**
- **Teaching Transparency, Action of Cocaine.** Use this transparency to illustrate a model of drug addiction. Point out the changes in the amounts of neurotransmitter in the synapse and in the number of receptor proteins. Have students explain why the neuron in part 3 has a reduced number of receptor proteins in its membrane. **(GENERAL)**
- **Teaching Transparency, Effects of Blood Alcohol Concentration.** Use this transparency to discuss the effects of alcohol at various concentrations in the blood. **(GENERAL)**
- **Group Activity**, Social Drug Use, TE. Have students bring in newspaper or magazine articles that deal with smoking cigarettes and drinking alcohol. Display the articles in the classroom. Ask students to consider the image of social drug use that is portrayed in the media. **(GENERAL)**

CLOSE *(15 minutes)*

- **Reteaching,** TE. Students work in pairs to write definitions for the key terms in this section and then quiz each other. **(BASIC)**
- **Quiz,** TE. Students answer questions that review the section material. **(GENERAL)**

HOMEWORK

- **Alternative Assessment,** TE. Students design an informative poster to educate people about the dangers of smoking. **(GENERAL)**
- **Section Review,** SE. Assign questions 1–6 for review, homework, or quiz. **(GENERAL)**
- **Science Skills Worksheet, CRF.** Students interpret data about nerve fibers and interpret a diagram of a synapse. **(GENERAL)**
- **Quiz, CRF.** This quiz consists of ten multiple choice and matching questions that review the section's main concepts. **(BASIC) Also in Spanish.**
- **Modified Worksheet, One-Stop Planner**. This worksheet has been specially modified to reach struggling students. **(BASIC)**
- **Critical Thinking Worksheet, CRF.** Students answer analogy-based questions that review the section's main concepts and vocabulary. **(ADVANCED)**

Optional Block

LAB *(45 minutes)*

_ **Exploration Lab,** Calculating Reaction Times, SE. Students investigate reaction times and design an experiment to investigate influences on reaction times. **(GENERAL)**

_ **Datasheets for In-Text Labs, Calculating Reaction Times, CRF.**

Other Resource Options

_ **Occupational Application Worksheet, Forensic Toxicologist, One-Stop Planner.** Students learn about the career of forensic toxicologist and complete a worksheet on common poisons. **(GENERAL)**

_ **Internet Connect.** Students can research Internet sources about Blood Alcohol Concentration with SciLinks Code HX4026.

_ **go.hrw.com.** For worksheets, videos, and other teaching aids related to this chapter, visit the HRW Web site and type in the keyword HX4 NRV.

_ **Biology Interactive Tutor CD-ROM,** Unit 1 Cell Transport and Homeostasis. Students watch animations and other visuals as the tutor explains cell transport and homeostais. Students assess their learning with interactive activities.

_ **CNN Student News.** Find the latest news, lesson plans, and activities related to important scientific events at **cnnstudentnews.com**.

Lesson Plan

End-of-Chapter Review and Assessment

Pacing

Regular Schedule: 2 days

Block Schedule: 1 day

KEY	
SE = Student Edition	TE = Teacher Edition
CRF = Chapter Resource File	

Block 10

REVIEW *(45 minutes)*

_ **Study Zone,** SE. Use the Study Zone to review the Key Concepts and Key Terms of the chapter and prepare students for the Performance Zone questions. **(GENERAL)**

_ **Performance Zone,** SE. Assign questions to review the material for this chapter. Use the assignment guide to customize review for sections covered. **(GENERAL)**

_ **Teaching Transparency, Concept Mapping.** Use this transparency to review the concept map for this chapter. **(GENERAL)**

Block 11

ASSESSMENT *(45 minutes)*

_ **Chapter Test, Nervous System, CRF.** This test contains 20 multiple choice and matching questions keyed to the chapter's objectives. **(GENERAL) Also in Spanish.**

_ **Chapter Test, Nervous System, CRF.** This test contains 25 questions of various formats, each keyed to the chapter's objectives. **(ADVANCED)**

_ **Modified Chapter Test, One-Stop Planner.** This test has been specially modified to reach struggling students. **(BASIC)**

Other Resource Options

_ **Vocabulary Review Worksheet, CRF.** Use this worksheet to review the chapter vocabulary. **(GENERAL) Also in Spanish.**

_ **Test Prep Pretest, CRF.** Use this pretest to review the main content of the chapter. Each question is keyed to a section objective. **(GENERAL) Also in Spanish.**

_ **Test Item Listing for ExamView® Test Generator, CRF.** Use the Test Item Listing to identify questions to use in a customized homework, quiz, or test.

_ **ExamView® Test Generator, One-Stop Planner.** Create a customized homework, quiz, or test using the HRW Test Generator program.

TEST ITEM LISTING
Nervous System

TRUE/FALSE

1. ____ The elongated extension of a neuron that conducts nerve impulses is called an axon.
Answer: True Difficulty: I Section: 1 Objective: 1

2. ____ The cells of the nervous system that conduct electrical signals are called nerve cells, or neurons.
Answer: True Difficulty: I Section: 1 Objective: 1

3. ____ Many neurons have a layer of insulation called a myelin sheath that wraps around the axon.
Answer: True Difficulty: I Section: 1 Objective: 1

4. ____ Myelin sheaths slow down nerve impulses.
Answer: False Difficulty: I Section: 1 Objective: 1

5. ____ The potential of a resting neuron is positive.
Answer: False Difficulty: I Section: 1 Objective: 2

6. ____ When a neuron is not conducting a nerve impulse, the neuron is said to be at rest.
Answer: True Difficulty: I Section: 1 Objective: 2

7. ____ Neurons communicate with other cells by using neurotransmitters at synapses.
Answer: True Difficulty: I Section: 1 Objective: 4

8. ____ Neurotransmitters are chemical messengers that carry nerve impulses across the synaptic cleft.
Answer: True Difficulty: I Section: 1 Objective: 4

9. ____ A neurotransmitter may either excite or inhibit the cell it stimulates.
Answer: True Difficulty: I Section: 1 Objective: 4

10. ____ The peripheral nervous system carries all the messages back and forth between the central nervous system and the rest of the body.
Answer: True Difficulty: I Section: 2 Objective: 1

11. ____ The spinal cord and the brain make up the peripheral nervous system.
Answer: False Difficulty: I Section: 2 Objective: 1

12. ____ Motor neurons that conduct impulses to skeletal muscles under our conscious control make up the limbic system.
Answer: False Difficulty: I Section: 2 Objective: 1

13. ____ Most of the activity of the cerebrum occurs in the cerebral cortex.
Answer: True Difficulty: I Section: 2 Objective: 2

14. ____ The spinal cord is a dense cable of nervous tissue that runs through the vertebral column.
Answer: True Difficulty: I Section: 2 Objective: 3

15. ____ The spinal cord links the brain to the central nervous system (CNS).
Answer: False Difficulty: I Section: 2 Objective: 3

16. ____ Some activity in the somatic nervous system, such as spinal reflexes, is involuntary.
Answer: True Difficulty: I Section: 2 Objective: 4

17. ____ A spinal reflex is a self-protective motor response.
Answer: True Difficulty: I Section: 2 Objective: 4

18. ____ Spinal reflexes are usually slow because they involve the spinal cord and the peripheral nervous system.
Answer: False Difficulty: I Section: 2 Objective: 4

19. ____ The peripheral motor neurons that regulate smooth muscles are part of the autonomic nervous system.
Answer: True Difficulty: I Section: 2 Objective: 5

20. ____ Emotions are controlled by the sympathetic division of the autonomic nervous system.
Answer: False Difficulty: I Section: 2 Objective: 5

21. ____ The sympathetic division decreases blood pressure, heart rate, and breathing rate.
Answer: False Difficulty: I Section: 2 Objective: 5

22. ____ Pain receptors are located throughout all tissues and organs except the brain.
Answer: True Difficulty: I Section: 3 Objective: 1

23. ____ Most of the pain receptors in the body are located in the brain.
Answer: False Difficulty: I Section: 3 Objective: 1

24. ____ Processing sites for sensory systems tend to be localized in specific regions of the cerebrum.
Answer: True Difficulty: I Section: 3 Objective: 2

25. ____ Rods are receptor cells that respond to bright light.
Answer: False Difficulty: I Section: 3 Objective: 3

26. ____ Most visual processing takes place in the temporal lobe.
Answer: False Difficulty: I Section: 3 Objective: 2

27. ____ Your ears help you maintain your balance.
Answer: True Difficulty: I Section: 3 Objective: 4

28. ____ We hear by detecting vibrations in the ground.
Answer: False Difficulty: I Section: 3 Objective: 4

29. ____ Taste buds located in the tongue are stimulated when a chemical dissolved in saliva binds to taste cells in the taste buds.
Answer: True Difficulty: I Section: 3 Objective: 5

30. ____ Chemicals in the air stimulate olfactory receptors.
Answer: True Difficulty: I Section: 3 Objective: 5

31. ____ The sense of smell affects the enjoyment of food.
Answer: True Difficulty: I Section: 3 Objective: 5

32. ____ Cocaine acts by causing dopamine reuptake.
Answer: False Difficulty: I Section: 4 Objective: 3

33. ____ Addiction to psychoactive drugs is a physiological response because addiction involves interactions of drug molecules with neurons and synapses.
Answer: True Difficulty: I Section: 4 Objective: 2

34. ____ Enkephalins are natural pain relievers released by the body in response to pain and stress.
Answer: True Difficulty: I Section: 4 Objective: 3

35. ____ Narcotics mimic the action of enkephalins.
Answer: True Difficulty: I Section: 4 Objective: 3

36. ____ Alcohol is able to change the structure of a neuron's membrane, changing the shape of receptor proteins.
Answer: True Difficulty: I Section: 4 Objective: 3

37. ____ Nicotine mimics the action of the neurotransmitter acetylcholine.
Answer: True Difficulty: I Section: 4 Objective: 3

MULTIPLE CHOICE

38. "Antennae" that extend from a neuron and that receive information from other cells are called
a. axons.
b. cell bodies.
c. synapses.
d. dendrites.
Answer: D Difficulty: I Section: 1 Objective: 1

39. Nodes of Ranvier
a. strengthen axons.
b. slow nerve impulses.
c. occur in malfunctioning axons.
d. are gaps in the myelin sheath.
Answer: D Difficulty: I Section: 1 Objective: 1

40. The myelin sheath
a. transmits impulses from one neuron to another.
b. insulates synapses.
c. nourishes neurons.
d. insulates axons.
Answer: D Difficulty: I Section: 1 Objective: 1

41. unmyelinated axon : slow nerve impulses ::
a. neuron : being composed of many axons
b. nerve impulse : not traveling through axons
c. myelinated axon : fast nerve impulses
d. dendrite : sending information
Answer: C Difficulty: II Section: 1 Objective: 1

42. synapse : two neurons ::
a. neuron : two cell bodies
b. cell body : two axons
c. synapse : cell body and axon
d. axon : cell body and synapse
Answer: D Difficulty: II Section: 1 Objective: 1

43. Which of the following statements about the resting potential of a neuron is true?
a. There are many times more sodium ions outside the neuron than inside.
b. Sodium ions are in balance inside and outside the neuron.
c. There are fewer potassium ions inside the neuron than outside.
d. There are equal amounts of potassium and sodium ions inside and outside the neuron.
Answer: A Difficulty: I Section: 1 Objective: 2

44. After an action potential, the sodium-potassium pump helps
a. rebuild axon fibers.
b. restore the resting potential.
c. cause a stimulus.
d. All of the above
Answer: B Difficulty: I Section: 1 Objective: 3

45. When a neuron is at rest,
a. sodium ions are most concentrated inside the cell.
b. potassium ions are most concentrated inside the cell.
c. the outside of the cell is negatively charged.
d. All of the above
Answer: B Difficulty: I Section: 1 Objective: 3

46. A nerve impulse
    a. moves from the inside to the outside of an axon.
    b. moves from the outside to the inside of an axon.
    c. is the movement of an action potential along an axon.
    d. moves slowly.

    Answer: C   Difficulty: I   Section: 1   Objective: 3

47. When an impulse moves down the axon,
    a. sodium ions first rush out of the cell.
    b. a small part of the axon momentarily reverses its polarity.
    c. the resting potential of the cell does not change.
    d. potassium ions are pumped into the axon.

    Answer: B   Difficulty: I   Section: 1   Objective: 3

48. Neurotransmitters are
    a. electrical impulses.
    b. found only in neurons with myelin sheaths.
    c. released at synapses.
    d. produced by muscles.

    Answer: C   Difficulty: I   Section: 1   Objective: 4

49. When neurotransmitters cross a synapse and open ion channels in the membrane of a postsynaptic neuron, which of the following happens?
    a. inhibition of impulses in the neuron
    b. the death of the neuron
    c. initiation of an impulse in the neuron
    d. Both (a) and (c)

    Answer: D   Difficulty: I   Section: 1   Objective: 4

50. When a neurotransmitter is released from a presynaptic neuron, the neurotransmitter may
    a. become an enzyme in the space between the neurons.
    b. bind to membrane receptor proteins on the membrane of the postsynaptic neuron.
    c. cover the membrane of the axon.
    d. cause the cell body of the postsynaptic neuron to enlarge.

    Answer: B   Difficulty: I   Section: 1   Objective: 4

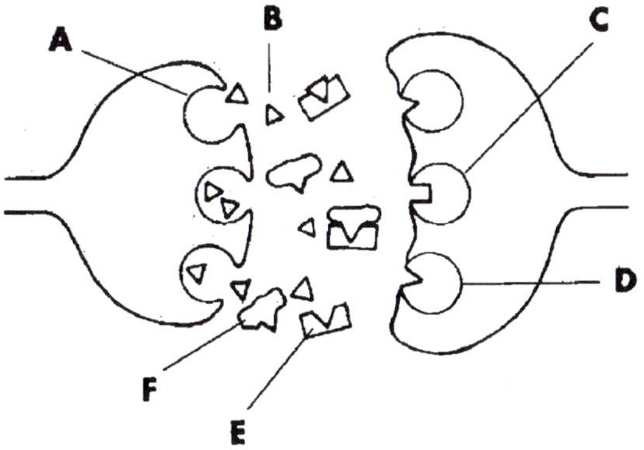

51. Refer to the illustration above. If neurotransmitters could not be cleared out of a synapse after transmitting a message,
    a. a postsynaptic neuron would continue to be stimulated for an indefinite period of time.
    b. the presynaptic neuron could not pass on its impulse.
    c. the postsynaptic neuron would not be stimulated.
    d. None of the above

    Answer: A Difficulty: II Section: 1 Objective: 4

52. Refer to the illustration above. Label B indicates a
    a. neurotransmitter molecule.
    b. neuromodulator molecule.
    c. receptor protein.
    d. drug molecule.

    Answer: A Difficulty: II Section: 1 Objective: 4

53. Refer to the illustration above. Which labeled object in the diagram would be responsible for removing neurotransmitters from the synaptic cleft?
    a. B
    b. C
    c. E
    d. F

    Answer: C Difficulty: II Section: 1 Objective: 4

54. Refer to the illustration above. The effect of the neurotransmitter might be prolonged by the presence of molecule
    a. B.
    b. D.
    c. E.
    d. F.

    Answer: D Difficulty: II Section: 4 Objective: 2

55. Refer to the illustration above. When a drug blocks removal of a neurotransmitter from a synaptic cleft for a prolonged period,
    a. the postsynaptic neuron is overstimulated.
    b. the number of receptors on the postsynaptic neuron decreases.
    c. the only way to maintain normal functioning of the nerve pathway is to continue taking the drug.
    d. All of the above

    Answer: D Difficulty: II Section: 4 Objective: 2

56. The central nervous system consists of
    a. the brain and spinal cord.
    b. spinal nerves only.
    c. the spinal cord only.
    d. the brain only.

    Answer: A Difficulty: I Section: 2 Objective: 1

57. The peripheral nervous system
    a. is not linked to the central nervous system.
    b. provides pathways to and from the central nervous system.
    c. consists of the cerebellum and spinal cord.
    d. is composed of only motor neurons.

    Answer: B Difficulty: I Section: 2 Objective: 1

58. Information is carried from the central nervous system to a muscle or gland by
    a. sensory neurons.
    b. sensory receptors.
    c. motor neurons.
    d. None of the above

    Answer: D Difficulty: I Section: 2 Objective: 1

59. Motor neurons transmit messages
    a. to the brain.
    b. to the spinal cord.
    c. from the spinal cord to the brain.
    d. from the central nervous system to a muscle or gland.

    Answer: D Difficulty: I Section: 2 Objective: 1

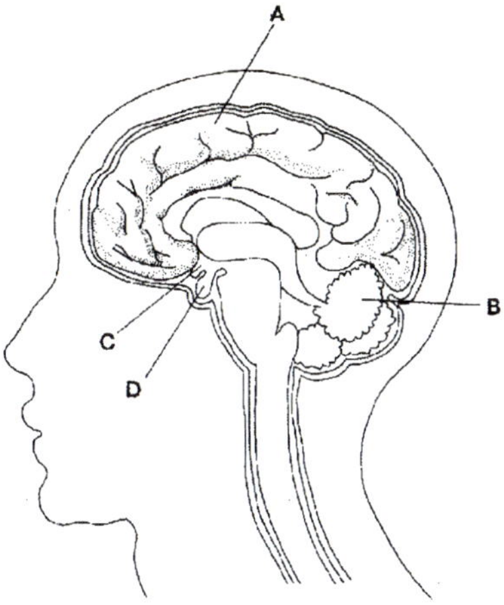

60. Refer to the illustration above. Structure B in the diagram is the
    a. spinal cord.
    b. brain stem.
    c. cerebellum.
    d. cerebrum.

    Answer: C   Difficulty: II   Section: 2   Objective: 2

61. Refer to the illustration above. Hunger and thirst are regulated by structure
    a. A.
    b. B.
    c. C.
    d. D.

    Answer: D   Difficulty: II   Section: 2   Objective: 2

62. Refer to the illustration above. The cerebrum is labeled
    a. A.
    b. B.
    c. C.
    d. D.

    Answer: A   Difficulty: II   Section: 2   Objective: 2

63. Sensory neurons transmit messages
    a. from the central nervous system to a muscle or gland.
    b. from the brain to the spinal cord.
    c. to the spinal cord or brain.
    d. All of the above

    Answer: C   Difficulty: I   Section: 2   Objective: 1

64. Gray matter consists of
    a. cell bodies of neurons.
    b. only synapses.
    c. myelin.
    d. axons.

    Answer: A   Difficulty: I   Section: 2   Objective: 3

65. The spinal cord is linked to the peripheral nervous system through
    a. spinal nerves
    b. the thalamus
    c. interneurons
    d. All of the above

    Answer: A   Difficulty: I   Section: 2   Objective: 3

66. A reflex
    a. is a sudden, involuntary movement of muscles in response to a stimulus.
    b. is not under conscious control.
    c. is not learned.
    d. All of the above

    Answer: D   Difficulty: I   Section: 2   Objective: 4

67. When the ligament below the patella is tapped, the quadriceps contracts, the hamstrings relax, and the leg rapidly
    a. contracts
    b. relaxes
    c. extends
    d. None of the above

    Answer: C Difficulty: I Section: 2 Objective: 4

68. The autonomic nervous system controls
    a. reflexes.
    b. voluntary movement.
    c. involuntary functions of the internal organs.
    d. locomotion.

    Answer: C Difficulty: I Section: 2 Objective: 5

69. The division of the autonomic nervous system that keeps you breathing when you fall asleep is the ____ division.
    a. somatic
    b. central
    c. sympathetic
    d. parasympathetic

    Answer: D Difficulty: I Section: 2 Objective: 5

70. Which of these is *not* a lobe of the brain?
    a. occipital
    b. parietal
    c. auditory
    d. temporal

    Answer: C Difficulty: I Section: 3 Objective: 2

71. How many lobes are there in the brain?
    a. 2
    b. 4
    c. 6
    d. 8

    Answer: B Difficulty: I Section: 3 Objective: 2

72. The layer of photoreceptors and neurons at the back of the eye is called the
    a. retina.
    b. cochlea.
    c. iris.
    d. optic nerve.

    Answer: A Difficulty: I Section: 3 Objective: 3

73. Dim-light vision is detected by the
    a. cones.
    b. lens.
    c. cornea.
    d. rods.

    Answer: D Difficulty: I Section: 3 Objective: 3

74. Sensory receptors essential for balance are located in the
    a. eyes.
    b. eardrum.
    c. cochlea of the inner ear.
    d. semicircular canals.

    Answer: D Difficulty: I Section: 3 Objective: 4

75. Hair cells in the semicircular canals detect
    a. motion of the head.
    b. loudness.
    c. light.
    d. sounds.

    Answer: A Difficulty: I Section: 3 Objective: 4

76. Ears
    a. function to detect sounds.
    b. help maintain your balance.
    c. detect only internal stimuli.
    d. Both (a) and (b)

    Answer: D Difficulty: I Section: 3 Objective: 4

77. Specialized receptors that enable hearing are found in the
    a. cornea.
    b. semicircular canals.
    c. cochlea.
    d. cerebellum.

    Answer: C Difficulty: I Section: 3 Objective: 4

78. When we hear,
    a. sound waves enter the ear canal and strike the eardrum.
    b. the membrane in the cochlea moves.
    c. the auditory nerve carries nerve impulses to the brain.
    d. All of the above

    Answer: D Difficulty: I Section: 3 Objective: 4

79. Chemoreceptors that detect odors are called ____ receptors.
    a. auditory
    b. olfactory
    c. cone
    d. cochlea

    Answer: B Difficulty: I Section: 3 Objective: 5

80. Psychoactive drugs, such as cocaine,
    a. affect the central nervous system by changing the activity of synapses.
    b. are usually not addictive.
    c. include only illegal drugs.
    d. All of the above

    Answer: A Difficulty: I Section: 4 Objective: 1

81. Cocaine acts by
    a. preventing dopamine reuptake.
    b. affecting the limbic system, causing euphoria.
    c. overstimulating postsynaptic neurons.
    d. All of the above

    Answer: D Difficulty: I Section: 4 Objective: 1

82. Narcotics affect the nervous system's control of pain perception by
    a. blocking dopamine reabsorption.
    b. mimicking cocaine.
    c. inhibiting dopamine production.
    d. mimicking enkephalins.

    Answer: D Difficulty: I Section: 4 Objective: 3

83. Nicotine in tobacco causes pleasurable feelings by
    a. blocking dopamine reabsorption.
    b. mimicking the action of acetylcholine.
    c. inhibiting a neurotransmitter similar to dopamine.
    d. mimicking enkephalins.

    Answer: B Difficulty: I Section: 4 Objective: 3

84. The effects of alcohol on the human body include
    a. changes to the cell membrane of nerve cells.
    b. altered transmission of nerve signals.
    c. impaired coordination.
    d. All of the above

    Answer: D Difficulty: I Section: 4 Objective: 3

COMPLETION

85. A(n) ____________________ is a cell that conducts electrical signals.

    Answer: neuron Difficulty: II Section: 1 Objective: 1

86. Cytoplasmic extensions called ____________________ allow a neuron to receive information simultaneously from many different sources.

    Answer: dendrites Difficulty: II Section: 1 Objective: 1

87. Some axons are surrounded by an insulating structure called a(n) ____________________ ____________________.

    Answer: myelin sheath Difficulty: II Section: 1 Objective: 1

88. The electrical charge across the membrane of a neuron is caused primarily by different concentrations of ____________________ and ____________________ ions inside and outside the cell.
Answer: sodium, potassium Difficulty: II Section: 1 Objective: 2

89. Messages are carried across synapses by signal molecules called ____________________.
Answer: neurotransmitters Difficulty: II Section: 1 Objective: 4

90. The junction of a neuron with another cell is called a(n) ____________________.
Answer: synapse Difficulty: II Section: 1 Objective: 4

91. The part of the nervous system that does not include the spinal cord and brain is known as the ____________________ nervous system.
Answer: peripheral Difficulty: II Section: 2 Objective: 1

92. Nerves that control breathing, swallowing, heartbeat, and the diameter of the blood vessels are found in the ____________________ ____________________.
Answer: brain stem Difficulty: II Section: 2 Objective: 2

93. The thalamus, hypothalamus, and cells deep within the cerebral cortex of the brain make up the ____________________ system, which helps regulate emotions.
Answer: limbic Difficulty: II Section: 2 Objective: 2

94. A sudden, involuntary movement in response to a stimulus is called a(n) ____________________.
Answer: reflex Difficulty: II Section: 2 Objective: 4

95. The ____________________ division of the autonomic nervous system regulates involuntary functions during routine conditions.
Answer: parasympathetic Difficulty: II Section: 2 Objective: 5

96. The two divisions of the autonomic nervous system are the parasympathetic division and the ____________________ division.
Answer: sympathetic Difficulty: II Section: 2 Objective: 5

97. Sensory receptors that respond to tissue damage are called ____________________ receptors.
Answer: pain Difficulty: II Section: 3 Objective: 1

98. Specialized neurons that detect sensory stimuli are called ____________________ receptors.
Answer: sensory Difficulty: II Section: 3 Objective: 1

99. Sensory neurons leading from each sense organ to the brain come together at a common region in the ____________________ ____________________.
Answer: cerebral cortex Difficulty: II Section: 3 Objective: 2

100. The ____________________ is the light-sensing portion of the eye.
Answer: retina Difficulty: II Section: 3 Objective: 3

101. When light enters the eye, it activates photoreceptors called ____________________ and ____________________.
Answer: rods, cones Difficulty: II Section: 3 Objective: 3

102. When light enters the eye, it first passes through the ____________________.
Answer: cornea Difficulty: II Section: 3 Objective: 3

103. Auditory processing takes place in the ____________________ lobe of the brain.
Answer: temporal Difficulty: II Section: 3 Objective: 3

104. The ____________________ is a small, snail-shaped structure of the inner ear lined with hair cells.

Answer: cochlea Difficulty: II Section: 3 Objective: 4

105. The specialized sensory receptors found in the cochlea are called ____________________ cells.

Answer: hair Difficulty: II Section: 3 Objective: 4

106. Olfactory receptors are located in the roof of the ____________________ passage.

Answer: nasal Difficulty: II Section: 3 Objective: 5

107. A(n) ____________________ ____________________ is a globular cluster of cells specialized to detect the four basic types of chemicals found in foods.

Answer: taste bud Difficulty: II Section: 3 Objective: 5

108. Psychoactive drugs that decrease the activity of the central nervous system are called ____________________.

Answer: depressants Difficulty: II Section: 4 Objective: 1

109. Exposure to large amounts of a psychoactive drug over a prolonged period of time may cause a change in the functioning of synapses. The result of such exposure is called ____________________.

Answer: addiction Difficulty: II Section: 4 Objective: 2

110. Neurotransmitters released by the human body in response to pain are called ____________________.

Answer: enkephalins Difficulty: II Section: 4 Objective: 3

PROBLEM

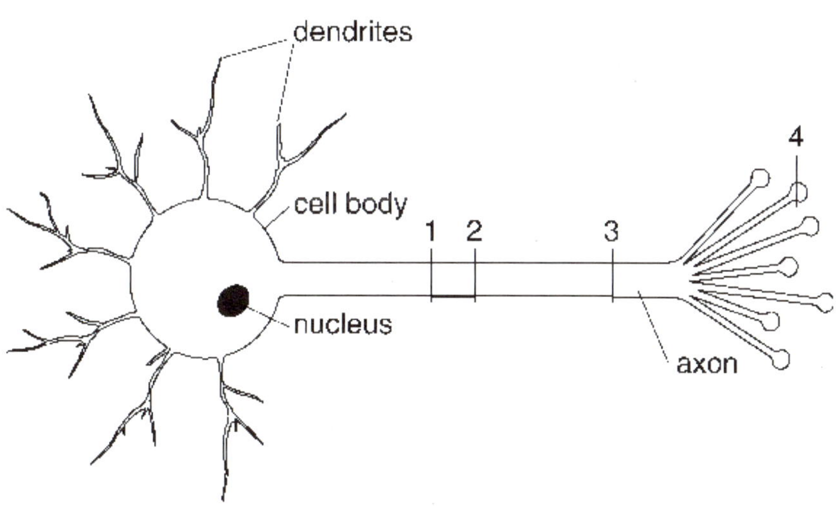

111. Refer to the illustration above. At the locations labeled 1, 3, and 4, the inside of the neuron is negatively charged compared to the outside. At the location labeled 2, the inside of the neuron is positively charged.
   a. Which numbered locations on the neuron are at the resting potential?
   b. Which numbered location on the neuron is showing an action potential?
   c. At which numbered location on the neuron is the cell membrane permeable to sodium ions?
   d. At which numbered locations on the neuron are sodium-potassium pumps actively pumping these ions?
   e. What happens when an action potential reaches location 4 on the neuron?

   Answer:
   a. 1, 3, and 4
   b. 2
   c. 2
   d. 1
   e. Neurotransmitters are released from the cell.

   Difficulty: III   Section: 1   Objective: 3

ESSAY

112. How is a signal from one neuron transmitted to another neuron?

   Answer:

   When a nerve impulse gets to the end of an axon, its message must cross the synaptic cleft. Messages are carried across synapses by chemical signal molecules called neurotransmitters. Neurotransmitters are found in vesicles at the end of axons. When a nerve impulse reaches an axon's end, the vesicles release neurotransmitters into the synaptic cleft. The neurotransmitters diffuse across the synaptic cleft and bind to receptors in the membrane of the postsynaptic cell. The postsynaptic cell is stimulated by the neurotransmitter molecules.

   Difficulty: III   Section: 1   Objective: 4

113. What part of the peripheral nervous system prevents you from holding your breath indefinitely? Explain.

   Answer:

   The autonomic nervous system carries messages to muscles and glands that usually work without our noticing, and enables the nervous system to maintain homeostasis within the body. We have involuntary control over some functions, such as breathing, that are regulated by the autonomic nervous system. Any voluntary control of the autonomic nervous system that endangers life disturbs the homeostasis of the brain tissue, causing unconsciousness. Then the autonomic nervous system takes over and restores normal functioning—in this case, breathing.

   Difficulty: III   Section: 2   Objective: 5

114. Explain why addiction to psychoactive drugs has a physiological basis.

   Answer:

   Some drugs cause excessive amounts of neurotransmitter to be present in synapses for long periods of time. This results in a decreased number of receptors on the postsynaptic membrane and a less sensitive nerve pathway. The only way a person who is addicted can maintain normal functioning of the nerve pathway is to continue taking the drug.

   Difficulty: III   Section: 4   Objective: 2

115. Describe the action of cocaine at the synapse and the effects of long-term use on receptors.

Answer:

Cocaine prevents the reabsorption of dopamine from the synaptic cleft. The trapped dopamine repeatedly stimulates postsynaptic neurons. The postsynaptic neurons adjust to the presence of cocaine by decreasing the number of dopamine receptors. This causes these neurons to become less sensitive, requiring more and more cocaine for stimulation.

Difficulty: III Section: 4 Objective: 2

9997266196 1 2 3 4 5 6